THE COSMOLOGICAL ORIGINS OF MYTH AND SYMBOL

FROM THE DOGON AND ANCIENT EGYPT TO INDIA, TIBET, AND CHINA

LAIRD SCRANTON

Inner Traditions
Rochester, Vermont • Toronto, Canada

Inner Traditions
One Park Street
Rochester, Vermont 05767
www.InnerTraditions.com

Library of Congress Cataloging-in-Publication Data
Scranton, Laird, 1953–
The cosmological origins of myth and symbol : from the Dogon and ancient Egypt to India, Tibet, and China / Laird Scranton.
p. cm.
Summary: "Reconstructs a theoretic parent cosmology that underlies ancient religion"—Provided by publisher.
Includes bibliographical references (p.   ) and index.
ISBN 978-1-59477-376-1
1. Mythology, Dogon. 2. Cosmology, Ancient. 3. Mythology—Comparative studies. I. Title.
BL2480.D6S37 2010
202'.409—dc22

2010019941

Printed and bound in the United States

10 9 8 7 6 5

Text design and layout by Virginia Scott Bowman
This book was typeset in Sabon with Schneidler Initials as the display typeface

To send correspondence to the author of this book, mail a first-class letter to the author c/o Inner Traditions • Bear & Company, One Park Street, Rochester, VT 05767, and we will forward the communication.

Cover figure (left): Male. Mali; Dogon (?). 12th–17th CE (?). Wood, 25⅜ x 3⅞ x 4⅝ in. (64.5 x 9.9 x 11.8 cm). Gift of Lester Wunderman, 1985 (1985.422.2). The Metropolitan Museum of Art, New York, NY, U.S.A. Image copyright © The Metropolitan Museum of Art / Source: Art Resource, NY.

CONTENTS

FOREWORD

From the moment I held Laird Scranton's first book *Sacred Symbols of the Dogon* I knew I was in for a mind-expanding excursion. In the foreword, John Anthony West spelled it out right from the start: "This book—small in size, large in significance—proves that the very latest scientific work on the structure and genesis of matter, quantum theory, and possibly both string theory and torsion theory was known in very ancient times. However, it was (and in certain cases still is) expressed in myth and symbol rather than in mathematical formulas."

Sacred Symbols of the Dogon united my left and right brains as it spiraled my imagination into an extraordinary and advanced mindset that existed on this planet millennia ago. I felt that no one has reached so far into the origin and meaning of symbols as Laird, and I was thrilled that he included us on the journey.

Laird's insights have brought clarity to questions many researchers have been asking as well as providing corroborating evidence of a lost, ancient "sacred" science based on a vastly detailed understanding of the basic forces of the universe.

For instance, I had made the uncanny match between the shape of the ancient Egyptian "ships of eternity" and modern depictions of wormholes. They're virtually identical symbolically and in stated purpose. I wondered if the gods of ancient Egypt shown riding upon these ships into the stars could have been opening wormholes and traveling through these cosmic tunnels or shortcuts to nearby star systems. Such a thought is, of course, nonsense to traditionalists. It blew my mind when Laird compared a diagram from string theory proponent Brian Greene's *The Elegant Universe*—of the fabric of space tearing and two ends of a wormhole starting to grow—with the ancient Egyptian hieroglyph for the word "to tear." They matched perfectly. I could feel Laird's excitement at this discovery, and share it. "Come closer," the gods seemed to be saying. "Open your minds, little ones. Reach for the stars. Sail the stars, with us."

Do wormholes exist in the universe? Are there ancient beings that can tear holes in space? Did they leave us clues for doing so, or better yet, a codebook containing the secrets of the nature of creation? These tantalizing questions stretch our minds to the limits. I, for one, am glad to have Laird Scranton doing the challenging and deep symbol work he is doing. His discoveries about this original symbol code have placed us on the cutting edge of knowledge and on the edge of a new world. Now it's up to us to set sail.

After reading Laird's work, who can read of a hole tearing open in heaven, as happened, for example, during the baptism of Jesus by John the Baptist (Mark 1: 9–11), and not wonder if there is some deep scientific message embedded in this crucial scripture. Before he was baptized, Mark refers to Jesus as "from Nazareth of Galilee" (Mark 1:9), whereas afterward he is "the Nazarene" (Mark 1:24), suggesting a transformation at the time of baptism. The Baptism of Jesus elevated him to a higher state of consciousness. Based on what Laird has revealed about ancient Egyptian myths and symbolism filtering into Judeo-Christian tradition, it is likely there is some secret, cosmological knowledge to be found here. We simply need to be brave and inquisitive enough to seek it.

Of late, it is becoming increasingly clear that it's going to take a revolution, or ascension, in our thinking in order to save our species, let alone fulfill our potential. It is going to take an integrated mind in tune with nature and her laws. This means a mindset that spends its precious energy on solving the mysteries of the universe (as opposed, for example, to contemplating the mysteries of why the economy won't improve or why this or that political party won't simply go away). This mindset is set on wholeness, even holiness. As Laird so beautifully tells us in his present work, this is the mindset of the ancients, and it leads to an "enlightening source."

The Cosmological Origins of Myth and Symbol extends the search for the "enlightening source" to the cosmologies of India, Tibet, and China. As Laird synchronizes these cosmologies with the ancient Egyptian and Dogon in his unique and elegant way, one readily realizes that what he has placed in our hands is a key to a legacy all of us have sought.

William Henry

William Henry is an investigative mythologist and author of numerous books on ancient mythology and symbolism. He is the host of the *Revelations* radio program.

ACKNOWLEDGMENTS

In addition to those already thanked in previous volumes, I am indebted to Rich Richins for the use of his beautiful image of Barnard's Loop, which was drawn from his website, www.enchantedskies.net. I would like to express my appreciation to Dr. Nilifur Clubwala, whose many inspired insights and insightful questions have led me again and again to push the boundaries of knowledge of my subject. I would again like to thank John Anthony West for his continued friendship and unwavering support. I would also like to thank Walter Cruttenden for the many kindnesses shown to me at his annual Conference on Precession and Ancient Knowledge. A very great debt of gratitude is owed to William Henry, who has championed my earlier books in his lectures and on radio, and for graciously agreeing to write the foreword for this book.

I would like to give many thanks to my nephew Matt Scranton and brother David Scranton, who were instrumental in acquiring a Na-khi–English dictionary for me in China. Thanks also to Christine Mathieu for her patience in allowing me to use her as a sounding board for my Dogon/Na-khi comparisons.

Thanks are due to the late Walter Fairservis, whose classes I unfortunately never took while at Vassar College, but who nonetheless eventually came to influence my outlook on the likely origins of ancient Chinese cosmology. I would again like to thank Nati Nataki at Afrikan World Books in Baltimore for his generosity and ongoing support of my work. I would like to thank Scott Creighton and Gary David for their insightful discussions, which helped to further the direction of this book. I also appreciate the steadfast interest and support that has been expressed for my books by our neighbors on Homestead Avenue in Albany. And last, but not least, I owe thanks to my wife, Risa, son, Isaac, and daughter, Hannah, for their quiet conspiracy throughout my studies to continue to treat me very much as they might treat an actual, normal person.

INTRODUCTION

If there were a single phrase or concept that could be said to best characterize ancient cosmology, it might be the phrase, originating in the Hermetic texts, "As above, so below." This statement expresses the cosmological idea of the "one thing"—the notion that what happens "above us" in the macrocosm of the universe is fundamentally similar to what transpires "below us" in the microcosm, at the next lower level of creation that conceptually precedes reality as we perceive it. Ancient concepts such as this have long held a fascination for the student of unsolved mysteries. No doubt this fascination is driven partly by our seemingly innate interest in anything old. By that standard, the myths and symbols of ancient cosmology, which are thought to date from the earliest dawn of antiquity, surely represent some of our oldest, and therefore most interesting and precious, artifacts. This book is the third in a series of volumes intended to shed new light on the nature of ancient cosmology and language. The field or discipline of ancient cosmology examines how these myths and symbols originally took their definition. The impulse to write these books comes as a response to the deep similarities in the cosmological myths of widespread cultures and as a reaction to the, in my view, largely inadequate rationales that are traditionally offered to explain these similarities.

The initial reaction of scholars to the many obvious parallels found among ancient cosmologies was, in my view, a sensible one:

they presumed that the cosmologies must be somehow related. During the eighteenth and nineteenth centuries, before the onset of modern theorizing on the subject, many individual attempts were made to promote one culture or another as an original seat of ancient cosmology, then trace its likely transmission to other regions through known migrations, invasions, and other cultural contacts. Ultimately, each of these proposed scenarios met with intractable contradictions or proved otherwise academically untenable, so no single culture could be designated as having invented the classic symbols and myths. The failure of this early approach to identify an origin for the cosmology set the stage for the introduction of other theories by which to explain the many persistent cosmological parallels that are evident from culture to culture. At least two such scenarios still retain popularity.

The first of these is the theory of polygenesis, the idea that cultures of similar levels of development and with access to comparable tools and materials will tend to create similar forms. My personal belief is that, while the concept of polygenesis may make complete sense, it is able to account for similarity of form only and offers little to explain the kinds of complex matching symbolism that often attend these forms. Consider the pyramid as an example. It seems reasonable that two distant cultures might each independently decide to stack large stone blocks together to create a permanent structure. On the other hand, it seems equally unreasonable that both would coincidentally choose to conceive the structure as a woman lying on her back, assign the same symbolic star groups to each of the four faces of the structure, and use the risings and settings of those star groups to govern their agricultural cycle. There is nothing to be found in the level of culture or in the simple availability of materials or tools to explain this. Surely some other influence must be at work here that goes beyond the simple idea of common cultural imperatives.

The second popular explanatory rationale, psychologist Carl

Jung's theory of archetypes and concept of a collective unconscious, suggests that mankind may have a genetic predisposition to formulate concepts of creation in specific, universal ways, that in essence the classic symbolic forms are "hardwired" into our psyche and that other, more subtle symbolic nuances of cosmology may have been communicated between cultures by way of a collective subconscious mind. In my opinion, the overwhelming problem with this view—beyond a very difficult burden of proof—is that it flatly ignores many well-documented statements by the ancient cultures themselves that again and again serve to define cosmology as a civilizing world plan and that credit its transmission to knowledgeable, revered ancestor/teachers. However, even if Jung's postulated concepts could somehow be shown to be true, the difference would be effectively transparent to our process; we would still be left substantially where we find ourselves—with an impulse to investigate the comparative contours of that original initial inbred cosmology.

If it seems likely that the ancient cosmologies cannot be just incidentally similar, and if we determine that these similarities cannot be reasonably assigned to the effects of polygenesis, then the presumption is that careful study may enable us, within the limits of the often fragmentary surviving evidence, to effectively synchronize the most markedly similar of these cosmologies by correlating their shared myths, mythological themes, symbols, deities, cosmological concepts, and cosmological words and meanings.

From this perspective, it is perhaps ironic that our inquiries into ancient myth and symbol, which are prerequisites to any attempt to synchronize these ancient cosmologies, began in the first volume of this series, *The Science of the Dogon,* not with discussions of some ancient culture, but rather with examination of the detailed cosmology of a modern-day African tribe from Mali called the Dogon. Although the Dogon are perhaps best known outside of cosmological circles for their artwork, which takes the form of slim wooden figures and carved wooden gate locks and granary doors, their religion is, in my opinion, central to ancient studies because it preserves, perhaps in more substantially complete form than that of any known ancient culture, many critical details of the cosmology. The Dogon religion constitutes an expansive symbolic system whose myths, mythological characters, deities, ritual acts, and cosmological words and drawings are cast in the familiar mold of the classic ancient cosmologies. In *The Science of the Dogon,* we illustrated many different ways in which key aspects of Dogon cosmology lend themselves to consistent correlation with various enigmatic elements of ancient Egyptian cosmology. We argued that the pervasive cultural, civic, and religious parallels that can be drawn between the Dogon and the ancient Egyptians strongly suggest a long period of close contact between the two cultures at some early point in Egyptian history and, as certain evidence suggests, specifically before the advent of written language. The reasonable presumption is that since we know with certainty that the system of culture in ancient Egypt retained much of its coherence for nearly three thousand years, then Dogon culture, which seems to have been founded on a markedly similar pattern, managed to sustain itself in coherent form for an additional two thousand years.

The Dogon priests define their cosmology, as symbolized by its ritual aligned granary, as a civilizing world plan, that is, a system of instructed civilization explicitly credited to ancient revered teachers.[1] The Dogon priests explain this plan in terms of a series of instructed Words (i.e., concepts or lessons), each of which defines an important civilizing skill. These range from the weaving of a cloth to the plowing of a field and the systematic organization of civic bodies. From this perspective, each aspect of Dogon culture that derives from the cosmological plan can understandably be seen to be somewhat reflective of it.

What we know about the Dogon religion comes primarily out of the decades-long studies of French anthropologists Marcel Griaule and Germaine Dieterlen, which began in the early 1930s and continued until the untimely death of Griaule in 1956. Griaule and Dieterlen documented a secret Dogon religion, an esoteric tradition that they said was known primarily to the Dogon priests and a handful of carefully screened initiates and whose inner details were largely unknown to the general Dogon populace. Griaule and Dieterlen reported what they learned about the Dogon tradition in two principal works. The first, titled *Dieu D'Eau* (the English edition was called *Conversations with Ogotemmeli*), records Griaule's first thirty-three days of instruction in this secret tradition by a Dogon priest. The second, Griaule and Dieterlen's finished report on the Dogon religion, which was completed by Dieterlen several years after the death of Griaule, is called *Le Renard Pale,* or *The Pale Fox.*

According to Griaule and Dieterlen, knowledge of Dogon cosmology is open to anyone who cares to actively pursue it.[2] In fact, according to Griaule, a Dogon priest is required to respond truthfully to any inquiry that is deemed appropriate to the questioner's initiated status and to remain silent—or even lie if pressed—in response to an inquiry that is deemed to be out of order or that exceeds the known initiated status of the questioner.

In this way, Dogon cosmology came to be associated with long-term study, driven first by the commitment of an individual to acquire it in an orderly way. Griaule's own experience as an initiate in the Dogon religion is reminiscent of the ancient Greek philosophers, some of whom reportedly studied with Egyptian priests for upward of twenty years before apparently coming back to Greece preaching concepts of creation expressed in classic terms that were, as Griaule himself noted, often markedly similar to those defined in the Dogon cosmology.

In 1975, the Dogon tribe came into controversy with the publication of Robert K. G. Temple's *The Sirius Mystery,* which focused on unexpected knowledge by the Dogon priests of obscure astronomic facts relating to the star system of Sirius—details that should not be detectable without the aid of a powerful telescope. According to Griaule and Dieterlen, the Dogon priests understood that Sirius is a binary star system, that is, one that gravitationally binds two stars together: a larger sunlike star (referred to by modern astronomers as Sirius A) and a much smaller, very dense dwarf star (called Sirius B.) Temple presented this information as evidence of a likely alien contact. However, since by the time these facts about Sirius were reported by Griaule and Dieterlen they had already been publicly documented by modern scientists, popular researcher Carl Sagan proposed that the Dogon knowledge was more likely the product of a modern-day contact with some outsider who was aware these facts, then integrated by the Dogon priests into the cosmology.[3] Dieterlen strongly disagreed with Sagan's suggestion and, as a way of countering it, produced a four-hundred-year-old Dogon artifact that featured a representation of the Sirius stars.

Several years later, in the 1980s, the Dogon were restudied by a second group of anthropologists led by Belgian Walter Van Beek. While it is not unusual for one anthropologist to restudy some aspect of the work of another, it *is* unusual for a scholar like Professor Van Beek, whose specialty had long been in the field of ecology, not religion, to presume to conduct such a comprehensive review of another anthropologist's major life work—work that had brought Griaule considerable fame in anthropological circles in France and beyond.

Notwithstanding Griaule's characterization of Dogon cosmology as a closely held secret tradition, these later researchers reported an inability to recreate Griaule's findings. Then, rather than concluding that the team had somehow simply failed to penetrate the secret

Dogon tradition, Van Beek inexplicably declared Griaule's Dogon cosmology to be a fabrication, one invented by the Dogon priests to satisfy the many persistent questions of Griaule. Years later, in a 2007 e-mail addressed to me personally, Professor Van Beek, who was initially concerned that I had based my studies on what he called the "quicksand of Griaule's cosmology," declared the Dogon granary, an aligned ritual structure that is central to Griaule's cosmology, to be a "chimera known only to Griaule." In his book *Dogon: Africa's People of the Cliffs,* Van Beek, so outwardly dismissive of his famous predecessor that he disdains to even mention Griaule by name, writes, "The Dogon have no creation myth, no deep story relating how the world came into being. (An anthropologist some decades ago probed his informants for creation myths so insistently that the Dogon, polite as ever, obligingly produced them. Some of his publications still in print as tourist guides perpetuate the mistake.)"[4]

However, unknown to both Griaule and his team, and apparently to all other Dogon researchers who followed during the next half-century after Griaule's death, was a critical fact relating to Dogon cosmology, one that might only elicit the attention of a comparative cosmologist and that stands in direct contradiction to Van Beek's view: the symbolism of Griaule's Dogon cosmology as it is evoked by the granary structure runs directly parallel to Buddhist cosmological symbolism as it is evoked by a *stupa,* a very similarly aligned ritual shrine. I first became aware of this fact in July 2005, when my daughter, Hannah, returned home from an educational trip to India with the Himalayan Health Exchange. She told me about a type of ritual shrine called a *stupa,* or *chorten,* that she had encountered throughout India, which struck her as being outwardly similar to the Dogon granary. Based on superficial resemblances between the two structures, I proposed to her that I could predict—sight unseen—several key aspects of symbolism that I felt were likely to be associated with a stupa. I made my list, which included alignment of the structure to the cardinal points, a base that symbolized the sun, a roof that symbolized the sky, and broad symbolic associations of both a biological and a cosmological nature, and she and I checked them together, using online search engines to validate our suspicions. Each of my predictions proved to be correct.

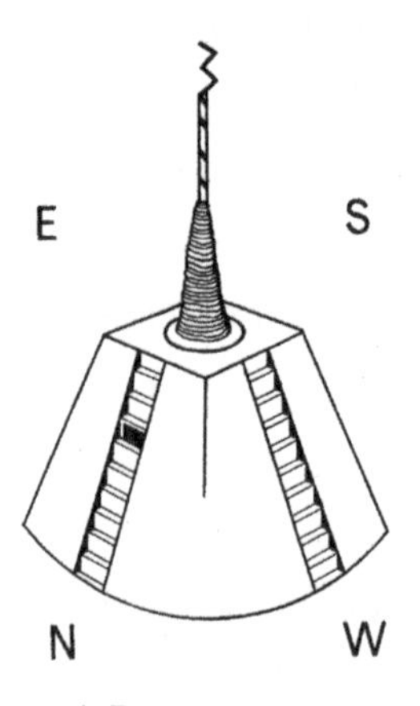

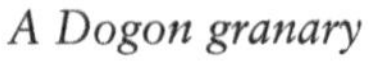
A Dogon granary

A Buddhist stupa

Next, I ordered a copy of *The Symbolism of the Stupa,* a text published in 1992 by Adrian Snodgrass of the University of West Sydney, Australia. Snodgrass is widely regarded as a leading authority on Buddhist architecture and symbolism. When I received Snodgrass's book, it was no surprise to learn that the Buddhist symbolism independently affirmed (strictly in terms of Buddhism and without regard to Griaule or his Dogon cosmology) virtually every key aspect of Griaule's Dogon cosmology. In other words, Dogon cosmology constitutes a known, legitimate cosmological form, one that surely could not have been just casually concocted by a group of Dogon priests, as Professor Van Beek suggests. The Dogon granary—Van Beek's "chimera"—is in all practicality a variety of stupa, an ancient ritual structure that is, in fact, widely known to Buddhists throughout India and Asia.[5]

Realizing the potential importance of the discovery, I responded to Professor Van Beek, both to bring the new information to his attention and to offer to co-report the new finding with him in *Current Anthropology*, the academic journal that he edited, which is published by the University of Chicago. To my disappointment, he never responded to my offer. A few months later I published an article on the subject under my name alone in *Anthropology News*, a second academic journal, also published by the University of Chicago.[6]

Dogon parallels to Buddhism are significant because they affirm the likely legitimacy of the Dogon esoteric system and uphold a great many of its details, precisely as Griaule reported them. As part of a cosmology whose form has been effectively affirmed by an independent source, the many direct parallels suggest that we can use Buddhist stupa symbolism as a fundamental basis of comparison and synchronization to that of the Dogon granary.

Griaule and Dieterlen state that *The Pale Fox* includes insights from priests of a number of tribes other than the Dogon who share the same fundamental cosmology.[7] Therefore, when appropriate, Griaule and Dieterlen chose to include minority opinions relating to a given concept or myth to produce a true synthesis of priestly opinion on the subject. Likewise, when discussing synchronized ancient cosmologies such as those relating to the Buddhist stupa and Dogon granary, we will see that different cultures may emphasize different aspects of what appears to be a single organizational plan. Where differences do arise, I will strive to credit each culture with its significant contributions to our understanding of that broader plan.

Based on the many likely relationships we have outlined in previous volumes between Dogon cosmological words and symbols and Egyptian hieroglyphic words and glyphs, one of the key resources we will turn to for reference in our study of comparative cosmology is the Egyptian hieroglyphic language itself. Likewise, because of what might be characterized as an almost predictive agreement between Dogon word pronunciations and meanings and those presented in Sir E. A. Wallis Budge's *An Egyptian Hieroglyphic Dictionary,* we have chosen to offer Dogon cosmological words as new evidence in support of Budge's much-maligned dictionary. Many modern Egyptologists have expressed disagreement with Budge's scholarship and feel that Budge's dictionary is out-of- date and that it can, at times, be somewhat unreliable. However, for a document that, in the traditional view, must be substantially wrong, Budge's dictionary has inexplicably shown itself to be a very close and consistent match for Dogon cosmological words and meanings—something it should not reasonably be able to do if Budge's view is fundamentally flawed. Consequently, Budge's dictionary remains my preferred choice for Dogon and Egyptian word comparisons. Another key resource for our study when comparing Dogon and Egyptian word meanings is a French dictionary of the Dogon language called *Dictionnaire Dogon,* which was compiled by Genevieve Calame-Griaule, daughter of Marcel Griaule and a well-respected anthropologist in her own right.

Our approach to reading Egyptian hieroglyphic words—the evolution and application of which was first introduced in *The Science of the Dogon* and documented extensively in *Sacred Symbols of the Dogon*—is a decidedly unorthodox one that begins *not* with the comparative texts of the Rosetta stone, but rather with cosmological symbols and words as they are defined by the Dogon priests. From this perspective, key shapes and meanings that originate with the Dogon and Buddhist cosmologies are applied to written Egyptian words simply by substituting well-defined cosmological concepts for associated shapes. In practice, each Egyptian cosmological word is treated as a symbolic sentence whose component glyphs define its meaning. For the purposes of these studies, when interpreting an Egyptian word, we read the glyphs from left to right, working from top to bottom as if in columns:

1	3	5
2	4	

One tangential purpose of the books in this series is to provide ongoing examples of Egyptian hieroglyphic words whose traditional meanings seem to be clarified through this type of straightforward symbolic interpretation.

Before we begin studying specific Egyptian word examples, we should quickly review key aspects of an unorthodox method of reading Egyptian hieroglyphic words—in particular, Egyptian cosmological words—that is first fully explained in *Sacred Symbols of the Dogon.* Within the mind-set of this method, a concept associated each glyph is substituted for the glyph when interpreting an Egyptian hieroglyphic word, much like a child's rebus puzzle. The meaning of the resulting symbolic sentence provides a definition for the written word. Likewise, there are some Egyptian hieroglyphic words that exhibit one or more unpronounced trailing glyphs whose meanings seem to relate to the word. Among traditional Egyptologists, the purpose of these glyphs is poorly understood. From the perspective of our unorthodox reading method, these words constitute *defining words,* whose purpose is to establish the symbolic meaning of the trailing glyph.

When attempting to read these words, it is important to understand that—even within the traditional view—an Egyptian glyph can carry more than one symbolic meaning. Likewise, it is a common attribute of Dogon and Buddhist cosmological words that they also carry more than one meaning. These meanings play out in Budge's hieroglyphic dictionary as homonyms—words that are pronounced the same way but carry different meanings.

Because I have already provided specific justification for the assignment of certain meanings or concepts to certain Egyptian glyphs,[8] it will not be my goal in this volume to rejustify every glyph usage. I will, however, strive to justify any glyph assignments that are new to this book and hope that the reader will, on that basis, take as justified the meanings I correlate to other Egyptian glyphs.

The esoteric nature of ancient cosmology is something of an enigma, in and of itself. If we are meant to interpret ancient cosmology as a plan of civilizing instruction, then we can only conclude that its details were organized and presented for our benefit and for the benefit of humanity in its entirety. So what was to be gained by hiding this information behind an apparent veil of secrecy? The Dogon priests make it clear that the esoteric tradition was meant to be open to any man or woman who might choose to pursue it in an orderly way. So the answer to a pivotal question is uncertain and remains for us to explore: from whom was the ancient esoteric tradition meant to be hidden?

In the end, the challenge for any proposed theory of ancient cosmology is that it explain, in believable terms, the themes, symbols, and constructs that have long been known to compose it but that have, for an equally long period of time, somehow eluded clear interpretation. It is the end purpose of this series of books to plumb the depths of ancient cosmology with hopes of returning with a clearer understanding of its likely purposes. Perhaps the best test of such an inquiry into the nature of ancient cosmology would be an ability to provide and support a credible interpretation for its signature phrase, "As above, so below."

ONE

CONCEPTS OF COMPARATIVE COSMOLOGY

By contemporary definition, cosmology is the study of the creation of the universe, the physical foundations of time and space, and the formation and structure of matter. Since the mid-1600s, the study of cosmology has been the near-exclusive domain of the astrophysicists—the Sir Isaac Newtons, Albert Einsteins, and Stephen Hawkings of the world—so it falls under the modern conceptual umbrella of science as opposed to religion. Ancient cosmology, on the other hand, along with other modern scientific disciplines like astronomy and mathematics, was for many thousands of years the traditional domain of the priests, so at the beginning of human civilization, it was effectively indistinguishable from religion. Although this difference may seem like a relatively minor one, in my opinion the modern expectation that there has always been a philosophical opposition between science and religion reflects a significant cultural filter through which we tend to view ancient cosmology and a major obstacle to a correct understanding of it.

If the earliest notions of creation had simply arisen unconstrained within ancient cultures, as is the popular belief, such that each culture came to establish its own somewhat quaint, ethnocentric view of how creation may have occurred, then today there might well not be any such thing as the study of comparative cosmology as we know it. In such a case, we would likely find no common basis for aligning the life view of, say, a native of the Amazon jungle with the qualitatively different experiences of an Easter Islander, or with those of an aborigine living in the Australian outback. Each culture might well have come to explain creation in its own terms, related to its own local circumstances and environment. However, when we study the actual creation traditions of distant cultures, uniqueness of view is not what we typically find. Rather, what we see instead is an almost predictive commonality of theme, symbol, and storyline, expressed in distinctly similar terms and organized according to a set of familiar stages of creation.

In the view of the comparative cosmologist, any credible approach to interpreting the many worldwide parallels in ancient cosmology must begin with the explicit statements of the ancient cultures themselves. As we have suggested, in culture after culture, these often reflect a vision of ancient cosmology as a kind of instructed system of civilization, one that was typically associated with knowledgeable ancestor/teachers or beneficent ancestor/gods. This explicit view of cosmology, which has most often been effectively set aside and disregarded by traditional researchers, is actually consistent with a large body of additional evidence that has been documented for these same cultures and that may offer us a coherent explanation for the many commonalities of ancient myth and cosmology.

The notion of an instructed civilization might explain the seemingly abrupt appearance of sophisticated cultural developments in the earliest societies, such as the highly refined work in pottery from the early days of ancient Egypt, the examples of ancient skilled master stonework that survive worldwide, and the sophisticated symbolic systems of writing that have been uncovered in Egypt, China, Central America, and elsewhere.

This alternate perspective, from which ancient cosmology is interpreted as an instructed system, does not rely on postulated psychology for its efficacy, but rather suggests that the common threads of so many world cosmologies imply a common parent cosmology. Perhaps the simplest and most cogent explanation for similar widespread systems of symbol and myth would be to argue that, somewhere in deepest antiquity, many of them may have shared a common parent.

The study of comparative cosmology provides us with a methodology for exploring the likely contours of this proposed parent cosmology, one that is perhaps best explained by way of a simple analogy. Imagine for a moment that you are a babysitter who is responsible for the care of twin toddlers each afternoon after picking them up from a morning daycare program. You know that the parents of the toddlers prefer to dress them identically, so they usually send them off in the morning wearing matching outfits. However, on this particular day, by the time you arrive at the daycare center, each child has managed to misplace several important articles of clothing. You realize that even though you never saw the children fully dressed that morning, it is still your job to somehow determine which articles of clothing have been mislaid by each child and retrieve them all before returning the children home again.

Given the likelihood that the outfits were originally identical, you decide that the best approach is to reconstruct the details of one full outfit by carefully comparing the toddlers' remaining attire. If one of the twins is missing his shoes, which he calls "booties," you realize that you can establish the specific type of shoe by simply asking to examine the booties of the other. If one is seen to be wearing a belt and the other is not, then you can assume that the second child's belt is missing. If one child is dressed in an undershirt, then you must verify that the other child is also still wearing one. If neither child has a sweater but one claims to be missing his, then it occurs to you that you can validate that claim by simply checking with the other child. Through these kinds of comparisons and corroborations, you come to see that a careful inventory of all of the items will improve your chances of returning home with most, if not all of the original items of clothing.

Much like the twin children in our analogy, the ancient cosmologies as they have been passed down to us often appear to be substantially incomplete. There are several reasons for this. First, our primary information about these cosmologies comes from written texts that, depending on the material on which they are written, can be fragile and may often have been found in incomplete form. Second, in some cases we may have recovered multiple copies of the same original text that are somewhat different from each other because of transcription errors or deliberate editing. Furthermore, some deities, myths, and symbols are known to have changed or evolved over time, perhaps as different cult centers or political figures came to worship different deities or emphasized different aspects of their own cosmologies.

For these reasons, the comparative cosmologist is often not in a position to directly compare symbolic elements on an apples-to-apples basis, and sometimes the differences between two cosmological systems can be substantial. For example, in ancient Egyptian cosmology, the great mother goddess, Neith (also known as Net), is credited with having woven matter on a loom with her shuttle, while in Dogon cosmology, the spider Dada (sometimes called Nana) is defined as having woven matter from threads in the form of web.[1] Although we might first see these as irreconcilable differences between the two cosmologies, when we dig deeper, we discover the likely truth of the matter. First, we find that in the Dogon language, the word *dada* actually means "mother," so it can be seen to agree with the Egyptian concept of a mother goddess. Second, when we look to ancient mother goddesses from other cultures that are traditionally equated with Neith, such as Athena in ancient Greece, we often find evidence of spider symbolism. These references help us to reconcile Neith in Egypt with the Dogon spider references. Then again, this same example demonstrates the importance of ultimate perseverance on the part of the comparative cosmologist, because when we eventually compare the Dogon and Egyptian cosmological systems with Buddhist cosmology, we find a tradition that defines matter as having been woven in a web by a spider as if on a loom with a shuttle.[2] In cases such as this, we find that with a broad enough base of evidence, some seemingly very large differences in the classic ancient cosmologies can often be discretely reconciled.

If we make the presumption that many of the world's cosmologies exhibit so many fundamental similarities because they were originally based on the plan of a common parent cosmology, then the study of comparative cosmology becomes a process of discovery that is essentially the same as our hypothetical inventory of the twin children's clothing. It begins with the selection of two or more outwardly similar cosmological systems and proceeds through careful examination of the key elements of each, taken as positive predictions of what should be found to be represented in the other. Just as the various specific articles of clothing become a target for comparison in our toddler analogy, comparative cosmologists are concerned with the similarities and differences in the cosmological themes, concepts, and symbols; the definitions and shapes of words; the names, titles, and traditional roles of important deities; and the defined interrelationships of those deities. Through these, the comparative cosmologist hopes to reconstruct the likely structural elements of the implied parent cosmology.

When discussing comparative cosmology, I find it helpful to think of each culture's system of cosmology as a kind of framework on which we can hang each component element. As part of our examination of any given element, we mentally place it in its appropriate location within the framework of the cosmology. Along with the element, we also hang any incidental references that might relate to it uniquely. For instance, we might place the Dogon creator god Amma at the top and center of the Dogon cosmological framework, along with references to Amma as a "hidden god," alternate definitions of the name Amma as meaning "to hold firmly, to embrace strongly and hold in the same place,"[3] the notion that Amma "holds the world between his two hands," and the Dogon concept of Amma as being "dual in nature."[4] We might also make note of the fact that, in the languages of some African tribes, the words *Amma* and *Amen* (*amen* also appears as an Egyptian word. The clearest comparison, however, is between the god names) are specifically equated.

Later, when we are ready to compare these Dogon cosmological elements with their likely counterparts from ancient Egypt, we might notice the Egyptian "hidden god" Amen hung in the same relative position on the framework of Egyptian cosmology, alongside the meaning of an Egyptian homonym pronounced "Amen" that means "to make firm, to establish" or "to fortify."[5] This framework approach helps us to conceptually organize the evidence and clearly illustrates the multiple points of agreement that may exist between two important cosmological symbols. (As an amusing side note, I once came across an Internet article whose author, apparently unaware that dual meanings are a signature characteristic of ancient cosmological words, stated that the Hebrew word *amen* could not be related to the Egyptian god Amen because it came from a root that meant "to establish.")

The use of the framework image as a conceptual tool is important in comparative cosmology because it provides another level of correspondence by which to help synchronize the cosmologies. We can illustrate the important role that the parallel frameworks play by citing another simple analogy. Think of these imaginary cosmological frameworks as an extended list of elements, similar to a list of the names of the months of the year. Also imagine that our task is to correlate the names of the months of the year in English with those in Spanish. We begin by simply placing each month name on a framework, in this case, by producing a written list of the names of the months in the order that they appear in a calendar:

NAMES OF THE CALENDAR MONTHS

ENGLISH	SPANISH
January	Enero
February	Febrero
March	Marzo
April	Abril
May	Mayo
June	Junio
July	Julio
August	Agosto
September	Septiembre
October	Octubre
November	Noviembre
December	Diciembre

When we review the two lists, we see that in many cases the English and Spanish month names (such as September and Septiembre) are so obviously similar that there can be no reasonable doubt of a correlation, even outside of the context of the matching lists. In other cases (like January and Enero) the words are less distinctly similar, so the correlation of the words might seem less obvious. However, in this particular case, our list does not consist of some random set of words from two languages, rather it represents two organized sets of words with a specific shared context—the months of the year—organized in matching sequence. Based on these criteria alone, we can argue that the two words *January* and *Enero* must positively correlate to each other because they hold matching positions within the two parallels lists that are known, themselves, to positively correlate. In the same way, the simple placement of a symbolic element within the larger context of its ordered cosmology becomes an important piece of information that we can cite as evidence when attempting to establish the likely correlation of similar cosmological elements.

Although the types of component elements that are examined by a comparative cosmologist may include cosmological words, it is important to emphasize the difference between these kinds of comparisons and formal linguistic analysis. It is *not* the intent of the comparative cosmologist to trace an etymological lineage for a given word in one culture to its likely counterpart in another. Rather, the intent is to demonstrate that the words hold comparable positions and meanings within the contexts of two parallel systems of cosmology. As evidence, the cosmologist may cite similarity of pronunciation, common multiple levels of meaning, or known relationships to other words, deities, or cosmological shapes or concepts. My contention is that the comparative cosmologist's efforts to correlate words in this way carry the same legitimacy as similar efforts related to other component elements of the cosmology. If the evidence is sufficient to suggest a likely correlation between the Dogon god Amma and the Egyptian god Amen, then the very same quality of evidence suggests the very same type of likely correlation between the words *amma* and *amen*. These are component elements of the Dogon and Egyptian cosmologies that, in all likelihood, correspond.

Working with ancient cosmology can also be very much like working with a toddler, often for similar reasons. Just as it is not always clear to a childcare worker what a toddler may be trying to say, it also is often not clear to researchers of cosmology what, if anything, the mythological symbols and storylines may have been meant to convey, so any progress in understanding often rests fundamentally on our ability to interpret correctly. Consequently, a good comparative cosmologist must take great care when proposing an interpretation and must also be prepared to specifically support and defend any interpretation.

In my experience, the most defensible interpretations relating to ancient cosmology are those that begin with an explicit, well-documented statement that has been drawn from the culture itself. Better yet is a well-documented statement that is specifically corroborated by equally direct statements from at least two different cultures in regard to some parallel aspect of similar cosmologies. For example, in the previous books of this series we have argued that ancient cosmology may ultimately serve to define scientifically accurate cosmological science. We adopted this approach based on direct statements of the Dogon priests, who believe that their cosmology describes how a tribal god named Amma created matter. Having begun with an unequivocal statement, the development of the broader interpretation relating the Dogon mythological structure of matter to the scientific structure of matter consisted simply of presenting the well-defined Dogon mythological components of matter side by side with their likely scientific counterparts and allowing the reader to simply observe the match. Later, it became clear that Buddhist stupa symbolism defines a matching structure that is also thought to define cosmological creation, using the same shapes attached to the same symbolic meanings. One key advantage to this approach is that it specifically limits the opportunities for wishful or speculative interpretation on the part of the researcher by placing the initial interpretive statement in the hands of the culture being studied.

From within this same mind-set, it is important for the comparative cosmologist to understand the critical difference between proving a point and demonstrating it. While planning my first book, I realized that in all likelihood there would be no argument I could conceivably present—no matter how cogent or well formulated—that could convince most readers that ancient myths could be a likely representation of correct science. On the other hand, it occurred to me that these same readers would likely find it difficult to muster an effective rebuttal to a direct, side-by-side comparison of myth and science couched in simple, parallel statements and drawings.

The notion of ancient cosmology as a designed system of civilizing instruction suggests that we should be able to apply a certain logic to our analysis of that system. For example, in a designed system we would expect the choice of symbols and the assignment of symbolism that relates to those symbols to make ultimate sense. Likewise, in a designed system, it should be possible to eventually understand what I call the mind-set of the designer, and from that understanding get a feel for how the planner of the system typically does things and make certain predictions based on predisposition or habit. Likewise, in a designed system one can presume that apparent patterns that recur within the system have intended meaning. All of this comes to be of great importance to a comparative cosmologist who is open to the idea that the larger system of ancient cosmology will make ultimate sense. Likewise, if we are to see ancient cosmology as a sensible system, then it becomes important for the researcher to take explicit statements at face value since these are the statements that fundamentally define the key elements of the system. Consequently, when the Dogon priests tell us that the *po* represents a primary component of matter, it would seem to be incumbent on us to consider the po in that context.

I am a convert to Judaism, and one of my long-standing complaints with the Jewish religion has always been the tendency of the great rabbis to treat as literal what I consider to be figurative statements (for example, I take the notion that the creation of the universe happened in seven days as a figurative statement) and literal statements figuratively. In my experience, perhaps the most critical skill for any researcher of ancient cosmologies to develop is a well-honed sense of when to simply take a statement at face value.

My professional training is as a business software developer, and it is largely because of my experience as the designer of sensible systems that I tend to notice what appear to be the designed aspects of ancient cosmology and language. Experience has taught me that a good software designer carefully considers how to make his system understandable to those who follow (perhaps even to himself at some later point in time when he may no longer be "in the mind-set" of the original system). One goal of good software design is to provide the next programmer with built-in clues to the meanings of the concepts and symbols that define a software program. Another is to try to make the system predictable to the next programmer by following consistent methods. Once we accept the proposition of ancient cosmology as a designed system, we begin to actively seek out various kinds of clues that might ultimately help us to understand the system. One purpose of the chapters that follow will be to define and document the specific kinds of clues that we believe are key to the plan of the ancient cosmologies.

TWO

SIGNATURE SIGNS OF THE PARENT COSMOLOGY

In previous volumes, we derived the likely attributes of my proposed parent cosmology through comparisons of the Dogon, Egyptian, and Buddhist cosmological systems. Since these are the likely attributes by which we will recognize the influences of this cosmology in other cultures, it makes sense that we begin by reviewing some of the key aspects of that cosmology. Many of these stand in unique relation to a very specific symbolic system, so their appearance can be taken as *signature signs* of the cosmology. By this, we mean that the mere appearance of these elements in an appropriate context within a given culture's creation tradition should alert us to the likely influence of the proposed parent cosmology.

Foremost among these signature signs are a few specific principles on which the parent cosmology is based. These principles express themselves in forms that should be familiar to the student of ancient religion because many of them also constitute classic themes and tenets of ancient mythology. These include the notion of creation from waves or water, the perception of universal principles of duality and the pairing of opposites, and the concept of processes of creation that are conceptually organized into discrete worlds or planes of existence.

Our referenced cosmologies agree that the universe began as a cosmogonic egg that held within it all of the unrealized potential of the future universe. Each tradition recognizes the terms *water, fire, wind,* and *earth* as referring to primordial elements or aspects of matter. Each recognizes the influence of a creator/god named similarly to Amma/Amen or Brahman. This god, who is closely associated with the primordial egg, effects creation by speaking a Word and is traditionally characterized as "hidden." Often, this creator/god is responsible for having introduced the concept of disorder into the universe as a consequence of performing an incestuous or masturbatory act. If we go far enough back in the formative stages of these traditions, we invariably find a mother goddess who has associations with spiders, a womb, storms, and the concept of weaving. Along with these, we also typically find quasi-historical associations with beneficent ancestor/teachers—perhaps themselves associated with the concept of an egg—who are credited with having introduced civilizing skills to humanity, such as weaving, agriculture, metallurgy, pottery, and written language.

Signature themes of the cosmology are often expressed in the folktales and myths of ancient cultures. As noted by Griaule himself, there are episodes in Dogon mythology that call to mind familiar storylines in classic Greek and Egyptian myths. These include ancient tales that describe the initial formation of a man from clay, relate how fire was stolen from the gods, and describe the firing of an arrow into the vault of the sky. Such themes—often expressed using images of serpents, clay pots, and spiraling coils—are all suggestive of the influence of the parent cosmology.

Part and parcel of this cosmological tradition are concepts that are expressed in relation to the numbers 7 and 8 and that are key to the cosmological symbolism as we understand it. These may relate closely to the concepts of death and life, respectively. Also associated are classical units for measuring time. These typically include the notion of a 30-day month and a 360-day/twelve-month year, a sixty-minute hour, and a sixty-second minute. Likewise we typically find the notion of the cubit as a unit for measuring distance, which is based on the concept of a bent arm and defined as the distance from the elbow to the tip of the middle finger.

The parent cosmology expresses itself in terms of cosmological concepts whose meanings are pivotal to a correct understanding of the cosmology. These concepts are often associated with specific cosmological shapes. In the Dogon and Egyptian traditions, these shapes/concepts are also associated with common cosmological words. Each word is typically assigned at least two distinct definitions that are characterized by a logical disconnect such that one meaning cannot be reasonably derived simply by knowing the other. An example of such a word is the word *Dogon* itself, which Griaule and Dieterlen tell us means "to complete the words" and "to remain silent." (Calame-Griaule lists the meaning of the word *dogo* as "to finish, to complete, to complete the words."[1]) These are the two implied obligations of an initiate in the Dogon esoteric tradition. The root of the word *Dogon* is *ogo,* which is also the name of a character who seems to play the role of light in Dogon myth. We take the word *ogo* as a likely correlate to the Egyptian word *aakhu,* meaning "light," and we know that Aakhu is also the name of the Egyptian god of light.[2] The same word also forms the root of the Dogon priestly title *hogon.* This apparent Dogon tradition of including the word for light in priestly titles is one that, according to North African ethnologist Helene Hagan, is also evident among the Amazigh. The Amazigh's name is a term that is applied to the predynastic hunter tribes who lived in Egypt prior to the First Dynasty, from whom the modern-day Berbers are thought to be descended and whose priestly titles also center on the Egyptian root *akh,* implying light.

Another example of a Dogon cosmological word with dual definitions is the word *po,* which refers to a primary component of matter and which comes from the same root as the word *polo,* which means "beginning."[3] These correspond in the Egyptian tradition to the word *pau,* meaning both "existence" and "primeval time," and Pau, the name of the Egyptian god of existence.[4] According to Budge's dictionary, the Egyptian word *pau-t* refers to "mass, matter, substance, the stuff of which all things are made." A word from the same Egyptian root refers to primordial time. According to Charles H. Long, a professor of the history of religion at the University of Chicago, a similar tradition exists among the Maori of New Zealand, in whose esoteric tradition the word *po* refers to "the stuff out of which and the matter by which creation comes into being"; the word can also refer to the time prior to the formation of the universe.[5] This view is upheld by Elsdon Best, who defines the Maori concept of the po in relation to the time prior to the formation of the universe and in regard to the coming into existence of the universe.[6] In the ancient Bon religion in Tibet, the word *ning-po* refers to "the essence," a cosmological term that is also applied to the stupa and that we take to mean "mass."

The importance of these discrete multiple meanings to a comparative cosmologist cannot be overstated because they represent a key factor by which the cosmologies of two cultures can be positively synchronized. From a strictly linguistic standpoint, two words of similar pronunciation and meaning may or may not imply a direct relationship between the words in question. In truth, given the relatively small number of phonetic sounds that are represented by spoken words, any apparent relationship between the pronunciations and meanings of words might too easily be the consequence of simple coincidence. However, when we introduce a second meaning into the equation in both languages, one that bears no logical relationship to the first except as defined within the somewhat obscure context of cosmology, it becomes possible to make a much stronger argument that the words must correlate, at least in terms of the cosmology—this notwithstanding the complaints of a professional linguist. When these same words can also be shown to relate to common symbols, deities, or cosmological themes, then in the eyes of a comparative cosmologist their correlation within the context of the cosmology is demonstrated almost beyond dispute.

As we have suggested, many of the Dogon cosmological key words are explicitly associated with at least one cosmological shape or cosmological drawing. In fact, Griaule and Dieterlen say that even though these drawings do not rise to the level of an actual written language among the Dogon, these drawn symbols—like ancient Egyptian glyphs—number in the thousands. Each cosmological key word defines a kind of conceptual package that stands uniquely on its own, consisting of a pronunciation, a cosmological concept, at least two well-defined and logically disconnected meanings, a related cosmological shape or drawing, and a specifically defined relationship to a stage of creation. My contention is that the concurrence of these same elements in combination with one another and taken in the context of another culture's outwardly similar cosmology constitutes an identity, a match so complete and precise that we can reasonably infer the influence of a commonly shared system of cosmology.

The Dogon priests refer to functional equivalences between certain civilizing acts, pairings that potentially serve the same kind of function as the logically disconnected meanings of the cosmological key words. For example, there is a direct comparison made between the acts of weaving a cloth and plowing a field.[7] Likewise, both the act of weaving a cloth and the act of cultivating a field are specifically equated with processes relating to the formation of a Word. The concept of the Word itself represents an act of speech that is closely associated with the processes that create life or that form matter. These are expressed in terms of the enigmatic notion that words are woven into the cloth, again both in relation to the formation of matter and in terms of human reproduction. The appearance of any of these phrases, concepts, or images in relation to the mythological tradition of another culture can be seen to signal the likely influence of our proposed parent cosmology.

Aligned structures are a very common feature of ancient cultures. Of these, many, like the Dogon granary and Buddhist stupa, were aligned to the four cardinal points of north, south, east, and west. Others were aligned to specific stars or star groups or to the rising or setting of these stars or star groups at a specific time of year, such as an equinox or solstice. Likewise, there has been much speculation regarding the specific methods that may have been used to align these structures to their various orientations. As regards the stupa, Snodgrass defines a very specific method of alignment that relies on the base plan of the structure for its function—a plan that is a close match for that of the Dogon granary.[8] The details of this method of alignment, when exhibited in another culture alongside other matching elements of the cosmology, become yet another signature sign of our parent cosmology.

Dogon cosmology as it is represented symbolically by the stupa/granary form evokes a series of shapes and associated meanings that are central to the cosmology and that constitute, in their own way, signature signs of the cosmology. These include the shape of the circle around a central dot and its associations with the sun, and the shape of a square associated with the concept of sky or space. Another such shape is that of a hemisphere or bubble associated with concepts of essence and substance or mass and matter. Relationships between these shapes and their assigned concepts can be taken as important indicators of the influence of the parent cosmology.

Although we might view Dogon cosmology in its entirety as a

grand symbolic construct that relates to the aligned ritual structure, there are specific symbolic assignments associated with individual structural elements of a Dogon granary that could also be taken as signatures of the cosmology. Among these are the agriculturally related star groups that are correlated to the faces of the granary. For the Dogon, these include the Pleiades, Orion, Venus, and an enigmatic star called "the long-tailed star."[9] The same can be said for specific symbolic references that relate to both cosmological and reproductive themes. When we encounter a ritual structure aligned to the cardinal points—one that is associated both with a womb and with concepts of the earth and sky—it seems reasonable for us to interpret it as signature evidence of our proposed parent cosmology.

There are aspects of early written language that we might also take as signature signs of the parent cosmology. Perhaps the most obvious of these are found in glyphs that take the same shapes as the series of well-defined figures that are evoked by the stupa/granary plan. When we see these same shapes employed in written hieroglyphic words to represent concepts that are a close match for those assigned within our parent cosmology, it is only reasonable to infer the likely influence of the cosmology. Likewise, we have seen how our symbolic readings of Egyptian hieroglyphic words reproduce meanings without regard to, or without the requirement for, vowel sounds. Consequently, when we encounter a language, such as ancient Hebrew, that traditionally omits written vowel sounds and yet is still somehow taken to have been primarily phonetic in nature, we can reasonably infer the influence of our cosmology.

The more deeply we delve into the realm of the Dogon initiate, the more we will recognize other themes, concepts, and symbols that are familiar to us from the study of world cosmology. These include many symbols and concepts that would seem to relate to astrology, including references to most of the animals and symbols of the zodiac and the concept of the axis and naval of the world, or *axis munde* and *umbilicus munde,* respectively. Likewise, we will see that the Dogon and Egyptian cosmologies together shed new light on the classic cosmological concept of the tree of life. Consequently, the appearance of any or all of these in relation to an ancient culture should spark our curiosity regarding a possible relationship to our proposed parent cosmology.

Ironically, what is perhaps the most obvious sign of our parent cosmology also seems to be the most universally ignored aspect of ancient studies. This is the direct statement, encountered again and again in culture after culture, that credits a group of honored ancestors/teachers—who were perhaps later deified—with the deliberate instruction of civilizing skills. It is not uncommon for these cultures to describe their ancient written languages as gifts of ancestors/gods. Such statements, which are routinely simply set aside without clear justification, seem to be taken by traditional researchers in the same vein and with the same shrugging lack of seriousness as a two-year-old's bedtime complaint that there is a boogeyman in the closet. We propose instead that these should be taken as signature aspects of the plan of the ancient cosmologies.

THREE

THE DOGON MYTHOLOGICAL STRUCTURE OF MATTER

According to the Dogon priests, one purpose of the key words and symbols of Dogon mythology is to describe how a tribal god named Amma created matter.[1] An understanding of the mythological structure of matter is important to our study of ancient cosmology because it provides a conceptual framework within which to interpret many of the diverse elements of the cosmology. One of my goals in the first volume of this series, *The Science of the Dogon,* was to present the component stages of the mythical Dogon structure of matter side by side with descriptions from popular modern astrophysicists such as Stephen Hawking and Brian Greene to demonstrate a likely relationship between Dogon myth and actual science.[2]

From the Dogon perspective, matter exists in three distinct Worlds,[3] first in a perfectly ordered state that is compared to waves of water. This well-ordered First World is disrupted by an act of perception that initiates a Second World of matter, where the concept of disorder is first introduced. Like the underworld in Egyptian cosmology, this disordering of matter is associated symbolically with the image of a jackal. From there, matter, now compared to primordial threads, is said to pass through a series of seven vibrations inside a tiny egg called the *po pilu,* or egg of the world, during which its structure is fundamentally reordered. This creates a Third World of matter, the world that we ultimately perceive. Consequently, the Dogon priests say that what we perceive constitutes only an image or reflection (albeit a correct image or reflection) of a more fundamental reality that exists in wavelike form far "below" our own plane of existence.

For the nontechnical layperson, who is likely to be familiar with the concept of atoms, protons, neutrons, and electrons, but may not be familiar with theories regarding the deeper workings of matter, these mythological stages of matter are perhaps best understood when they are given from the top down, starting with a basic atomlike unit of matter, which was mentioned previously, called the po. The Dogon priests state that "the things created by Amma will form themselves by the continuous addition of identical elements," beginning with the po.[4] According to the Dogon mythological model of matter, the po is composed of smaller components of matter called *sene* seeds that, like protons and neutrons in the nucleus of an atom, combine at the center of the po. Then the sene seeds, like an electron orbiting an atom, surround it by crossing in all directions to form a nest. The Dogon drawing that symbolizes this nest is a close match for the classic shape of an electron orbit as depicted from an electron microscope image.[5]

A likely counterpart in the Egyptian hieroglyphic language to the Dogon concept of the sene is expressed through the phonetic value *sen,* which carries several different sensible meanings that could be significant in terms of Dogon cosmology. These begin with an Egyptian glyph that is, unexpectedly, assigned two phonetic values, *sen* and *au* ⊂⊃.[6] The word *aun* means "to open"[7] and is written with the ✢ glyph, which we take as a likely counterpart to the Dogon nest drawing. This can be seen as the phonetic root of the word *aunnu,* which

Sene seed diagram

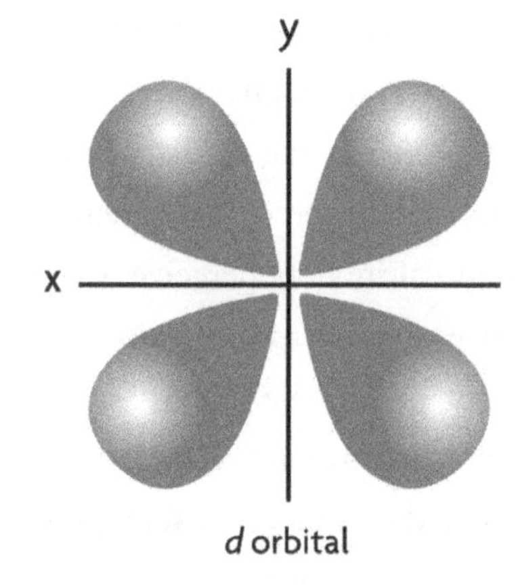

Electron orbital shape

means "nest."[8] There is also the Egyptian word *sen,* that means "to open."[9] Another Egyptian word, *senb,* means "to bind," which is the function served by both the sene seeds as they comingle at the center of the po, where they bind together like components of the nucleus of an atom, and by electrons in molecules, which are functionally shared to bind atoms together.[10]

According to the Dogon priests, the sene seeds are themselves composed of even smaller, more fundamental particles, which the Dogon define as the 266 primordial seeds or signs of matter. These are likely counterparts to the more than two hundred fundamental particles currently known to modern astrophysicists. Scientists organize these particles into four categories based on a property called spin that defines, based on rotational attributes of the particles, what each category of particle looks like from different perspectives. From a scientific perspective, some look the same from all angles (figure b page 35), some look the same if you rotate them halfway around (figure a page 35), some look the same if you rotate them all the way around (figure c page 35), and interestingly, some don't look the same unless you rotate them around *twice* (approximated by figure d below). The Dogon priests symbolize this aspect of their primordial seeds or signs using four circular drawings that, to the extent possible in two dimensions, exhibit the rotational attributes defined by these spin categories.[11]

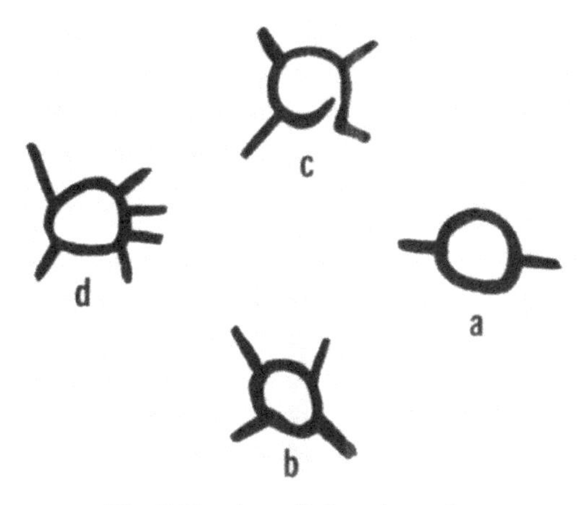

The 266 primordial seeds or signs

Like modern string theorists, the Dogon say that, conceptually, prior to existing as these particles, matter exists as primordial threads, which are effectively woven into matter. Each thread is said to pass through a series of seven vibrations inside a tiny egg, which the Dogon call the po pilu and which we take as a likely counterpart to the tiny, wrapped-up bundles of seven dimensions in string theory or torsion theory called the Calabi-Yau space. It is this component of matter that the Dogon priests call the egg of the world and describe as a pivotal component of matter to be found in the world just "below" ours. The vibrations inside this egg are conceived of as seven rays of a star of increasing length and are represented by yet another Dogon drawing. The figures of this drawing are read from right to left, like Egyptian glyphs as they are arranged in some inscriptions or like the letters of a traditional Hebrew text.[12]

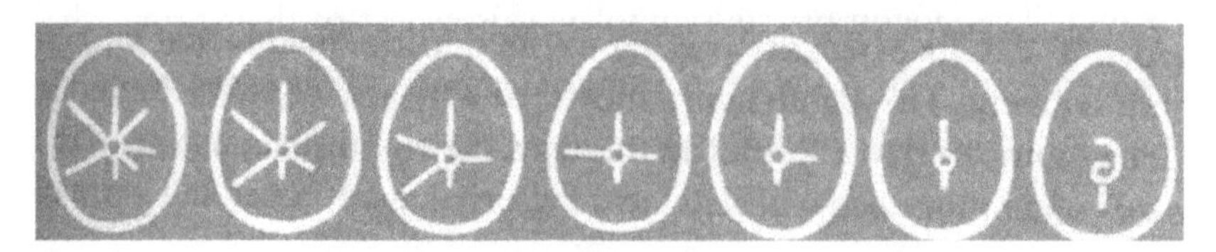

The seven stages of the po pilu

Not pictured in this drawing is an event that is considered by the Dogon priests to be an eighth stage of the po pilu, in which the seventh ray grows long enough to actually pierce the egg. This constitutes the conceptual death of the first egg and the act that is seen to initiate the birth of a new egg.[13] These eggs combine in a series—like pearls on a string, according to Buddhism—and are said to form membranes. These are likely counterparts to dimensional "branes" that form in string theory or torsion theory and are compared by the Dogon priests to the thin covering that surrounds the brain.[14] As a finished unit, the po pilu is characterized by the Dogon priests by the spiral that can be drawn to inscribe the endpoints of its seven rays. This spiral is a likely counterpart to a tiny vortex that is postulated in torsion theory to exist at this same level of matter.

Prior to the po pilu, the Dogon priests believe that matter exists in the form of waves, which are compared to waves of water and referred to by the term *nu*. These would correspond to massless waves in modern astrophysics and likely correlate to the Egyptian concept and name *Nun,* which refers to the deified concept of "the primeval water from whence everything came."[15] The same concept is likely represented by the Egyptian wave and waves glyphs, , which carry the pronunciations "n," "nu," or "mu."[16]

As we have suggested, the Dogon Second World—a likely counterpart to the Egyptian underworld—is a place where the concept of disorder (symbolized by the Dogon jackal) is first introduced to matter, and where matter is then effectively reordered into a new state.

This reordered state is symbolized in Dogon cosmology by an animal called the pale fox (*Vulpes pallida*),[17] a canine native to ancient Egypt whose job it is in Dogon cosmology to act as a judge between truth and error. Of course, the symbolic assignments of the jackal and the canid judge align well with the jackal of the Egyptian underworld, who is also understood to symbolize the concept of disorder, and the Egyptian jackal-headed god Anubis, who is traditionally interpreted as a judge between good and evil.

By comparison, the Egyptian underworld, the Tuat, is the conceptual counterpart to the Dogon Second World, and is traditionally considered a place of rebirth and is symbolized by a figure that features the rays of a star inside a circular egg . Discussions in the second volume of this series, *Sacred Symbols of the Dogon,* correlate key concepts of Dogon cosmology that relate to this egg to Egyptian words that express the same concepts in markedly similar terms. For example, an entry in Calame-Griaule's *Dictionnaire Dogon* indicates that the Dogon word *anu,* a likely counterpart to the phonetic root of the name of the Egyptian jackal-god Anubis, refers to "a seed that wraps around other seeds at the time of creation."[18] This is a functional description of the po pilu, which is a likely Dogon counterpart to the Egyptian Tuat.

Based on these correlations, we can see that the Egyptian underworld, which is traditionally interpreted as a place of the death and rebirth of souls, may in fact have been intended, as the Dogon myths suggest, to describe the disordering and reordering of matter in the microcosm. This interpretation would make sense, since the role of Neith, the great mother goddess of the Egyptian religion and mother of all of the Egyptian gods and goddesses, is explicitly defined in terms of the formation of matter.

By any standard, the mythological structure of matter as defined by the Dogon priests follows along in close parallel with the structure of matter as it is known to modern astrophysicists. A more detailed, side-by-side comparison of the two systems is presented in my book *The Science of the Dogon.* Likewise, the key Dogon cosmological words and shapes used to define this mythological structure are shown to be in specific agreement with ancient Egyptian hieroglyphic words and shapes as documented by Budge.

We can see, even based on the few examples discussed so far, that there seems to be an ongoing correspondence between Dogon cosmological drawings and Egyptian glyph shapes and concepts. We noted in the previous volumes of this series that these correspondences are so seemingly extensive as to allow us to outline in correct detail the same component stages of matter as described in Dogon and Egyptian cosmology, based on Egyptian glyph shapes and their apparent relationship to Dogon words and drawings.

FOUR

SYMBOLISM

Symbolism as we find it in Dogon society is a practice that has shown itself to be at least akin to writing in its ability to transmit essential knowledge in intact form from generation to generation. In that respect, we might think of symbolism as the effective language of ancient cosmology, the medium through which the meanings of cosmology were first recorded and by which they might later be replayed. Certainly, success in understanding key cosmological references often rests on our ability to meaningfully interpret ancient symbolism. Since the concept of symbolism plays such a pivotal role in our understanding of ancient cosmology, it became one of my first goals while preparing this manuscript to evolve a succinct definition for what constitutes the essential nature of a symbol. But somehow, the further I progressed with my studies, the more difficult this task seemed to become. For some time, I struggled to reconcile the many diverse types of symbolic references found in ancient cosmology, hoping to eventually "reduce fractions" and thereby arrive at some common denominator for what ultimately constitutes a symbol. But even after long deliberation, the best definitions that I could muster still seemed to fall somewhere short of the true mark.

Recently, I was reflecting on my previous Egyptian language studies, which suggest that Egyptian hieroglyphic words may often constitute symbolic definitions of the very concepts they convey. It occurred to me that, as one consequence of those studies, I had now gained insight into a seemingly unparalleled source for defining the central concepts of ancient cosmology—the Egyptian hieroglyphic dictionary itself. Moreover, the many Egyptian word examples cited, both in this volume and in previous volumes, suggest that Egyptian words can be relied on to provide direct, succinct symbolic explanations for otherwise obscure concepts. For that reason, I resolved to look to ancient Egyptian words relating to symbolism for a definitive statement regarding the essential nature of a symbol.

When, at last, I did search Budge's dictionary for the word *symbol,* what I found were two separate word entries, one pronounced "ashem" and the other "akham." I realized that *ashem,* or *Hashem,* was a word that I was already acquainted with based on familiarity with the Hebrew language; in Judaism it is a word that is substituted as a kind of placeholder (or symbol, if you will) for the name of God. (Often, when exploring the possible meanings of Egyptian cosmological words, I find it useful to compare words in the languages of other cultures that have similar cosmologies, such as the Dogon or Hebrew languages. Parallels between ancient Egyptian and Hebrew words occur with sufficient frequency that Budge often makes reference to a Hebrew word or letter as a way of explaining the likely pronunciation, meaning, or usage of an Egyptian word.)[1] This usage of the word *ashem* is consistent with that of a Dogon word *aduno,* which Calame-Griaule defines in her Dogon dictionary to mean "symbol"—as well as "world," "universe," or "creation."[2] *Aduno* is a cosmological term that I associate with the Hebrew word *adonai,* one that is also often substituted as a placeholder for the name of God in Judaism. Furthermore, I knew that the word *akham* seems to combine the omnipresent Egyptian prefix *akh,* which I associate with the concept of light or enlightenment,[3] with the suffix *am,* meaning "to know."[4]

A likely symbolic reading of the hieroglyphic word *ashem,* which Budge defines as "a symbol or figure of a god or sacred animal," rests on a view of the bent-arm glyph as representing an action or force, the divided river glyph as depicting the concept of a dam—such as might be used to create a reservoir—and the image of an owl in its familiar sense as a symbol for knowledge. Based on these definitions, the word *ashem* reads:

DEFINITION OF A SYMBOL (ASHEM)

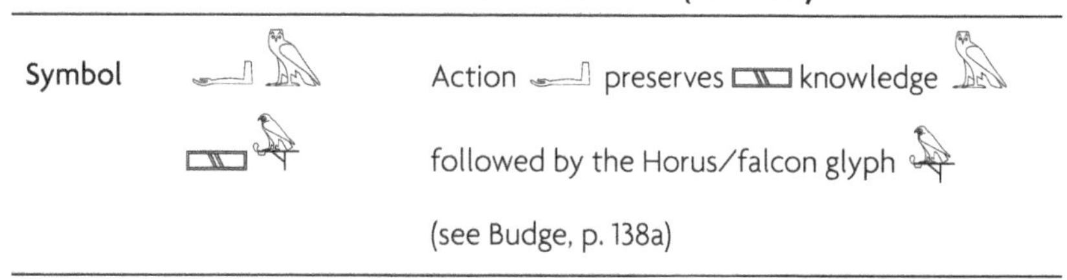

Symbol	Action preserves knowledge
	followed by the Horus/falcon glyph
	(see Budge, p. 138a)

I also noticed that the second word, *ahkam,* which Budge defines to mean "image or symbol of a god," follows almost precisely the same written form as the word *ashem,* but with the simple substitution of a figure Budge defines as a uterus for the divided river glyph. In Dogon cosmology, a uterus is seen as the vehicle for biological reproduction, so it symbolically implies the concept of replication or reproduction. Based on this interpretation, the word *akham* reads:

DEFINITION OF AN IMAGE OR SYMBOL (AKHAM)

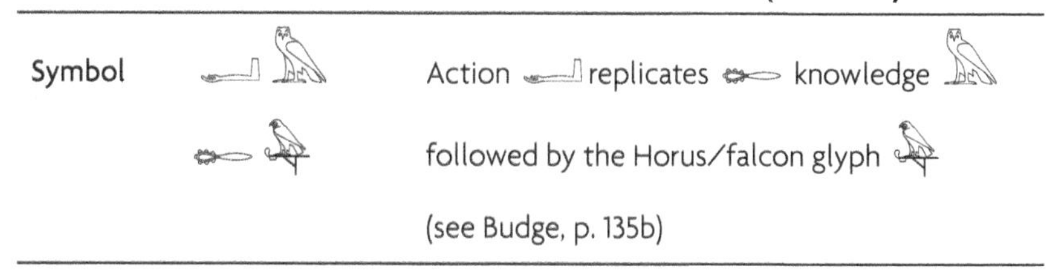

Symbol	Action replicates knowledge
	followed by the Horus/falcon glyph
	(see Budge, p. 135b)

These Egyptian word examples offer two significant perspectives on the concept of a symbol: one that seems given from the viewpoint of the person who creates or encodes the symbol, the other from the person who ultimately interprets it. Conceptually, they define a symbol in terms of a symbolic act that is intended first to preserve knowledge, then later to replicate it. Thus defined, a symbol becomes—as the meanings of the combined prefix and suffix *akh* and *am* suggest—a conduit for the transmission of "enlightening knowledge."

Given that in the modern view, symbols are thought of less in terms of actions and more in terms of discrete objects or images such as a drawn figure, a written character, or some ritually significant entity or event, it is interesting that these ancient Egyptian words would address the concept of symbolism first in terms of an act or action. Although this outlook may not resonate with a modern viewer, in Dogon society where no native written language exists and therefore no symbolic meaning can be said to have been preserved in written form, Griaule and Dieterlen tell us that the Dogon priests place similar emphasis on the importance of ritual acts when defining symbolism. In fact, each notable act of daily Dogon life is assigned cosmological symbolism, although, because of the esoteric nature of Dogon cosmology, the ultimate import of that symbolism may not always be fully grasped by the tribe member who actually performs the act. Thus, from the Dogon perspective, a natural and intimate relationship exists between ritual acts and cosmological symbolism.

Another observation to be made about the Egyptian words *ashem* and *akham* is that each displays the same trailing falcon glyph 𓅆, a figure that is traditionally associated in ancient Egypt with the god Horus.[5] If we hold true to the standards for interpreting Egyptian words that we first put forth in the second book of this series, *Sacred Symbols of the Dogon,* we realize that the commonality in the written form and meanings of these words, along with the common trailing glyph they share, signals the use of a convention; we see that these words take the form of defining words for the falcon glyph.

Furthermore, given the parallel written form of the words *ashem* and *akham* and the striking consonance of their symbolic readings, we might reasonably conclude that, on one level, the words were meant to assign the meaning of "symbol" to that glyph.

In support of our view of the relationship between ritual acts and the assignment of symbolic values to images, we can offer the Egyptian word *s-tut,* which Budge defines as meaning "to observe some custom, to do something that is usually done, to make a copy or an image, to fashion, to typify, to symbolize."[6] Symbolically, the word *s-tut* reads:

TO SYMBOLIZE (S-TUT)

To symbolize		To bind (a) matter with
		(the) matter's image
		(see Budge, p. 628b)

The written form of this word implies that just as it is the role of a symbolic ritual act to preserve and later recreate knowledge, it is also the role of symbolism to assign important matters or concepts to meaningful symbolic images. The more intimate the relationship between the image and its associated meaning, the more effective the symbol. The process by which images are evoked in relation to their assigned meanings plays a pivotal role in cosmology as the Dogon priests describe it.

The symbolic reading that we arrived at for the word *s-tut* above depends on an interpretation of the glyph as conveying the concept of an image, much as a statue is intended to reflect the image of its subject. This meaning is assigned to the glyph by another Egyptian hieroglyphic word, *ar,* or *aru,* meaning "form, figure, image, ceremony, rite."[7] The written form of the word and Budge's defined meaning again lead us to associate the notion of a symbol both with a ritual act and with the idea of a perceived image. The word *ar,* or *aru,* which can be construed as a defining word for the image glyph , is written:

DEFINITION OF AN IMAGE (AR OR ARU)

Image		That which is seen , followed by the image glyph (see Budge, p. 69a)

As we have mentioned in regard to the mythological structure of matter, according to both Dogon and Buddhist traditions, everyday reality as we perceive it constitutes a mere reflection or mirage, an image that is reflective of and ultimately derived from a more fundamental underlying reality. From this perspective, we can say that Dogon cosmology conceives of our familiar reality as a mere symbolic reflection. It is for this reason that, from the Dogon perspective, the concept of a symbol is considered fundamental. In fact, Griaule and Dieterlen tell us that for the Dogon priests, symbols are deemed ultimately more significant than the actual thing itself.[8]

The relationship that we suggest may have existed between symbolic ritual acts and the transmission of civilizing knowledge from some ancestral pedagogical source indicates that we should explore the concept of rites and rituals in ancient Egyptian language for further evidence that might support the concept. One Egyptian word meaning "to perform rites" is pronounced by Budge as "s-aakhu," and is based on the familiar root *aakhu,* an important, previously discussed cosmological word that we associate with the concept of light or enlightenment. Symbolically, the word reads:

TO PERFORM RITES (S-AAKHU)

To perform rites		To bind with enlightenment's source (see Budge, p. 589b)

This reading depends on an interpretation of the glyph as representing the concept of enlightenment. We can substantially affirm this interpretation by looking to a reference to the glyph in Budge's "List of Hieroglyphic Characters," which is found at the front of his *An Egyptian Hieroglyphic Dictionary.* Budge defines the glyph as symbolizing the concepts of "light, radiance, brilliance, shine," and offers a series of three glyphs to explain its phonetic value. However, we know based on discussions in *Sacred Symbols of the Dogon* that each phonetic value seems to relate to a cosmological concept. By substituting concepts for phonetic values, we interpret Budge's entry symbolically below:

LIGHT, RADIANCE, BRILLIANCE, SHINE (AAKH)

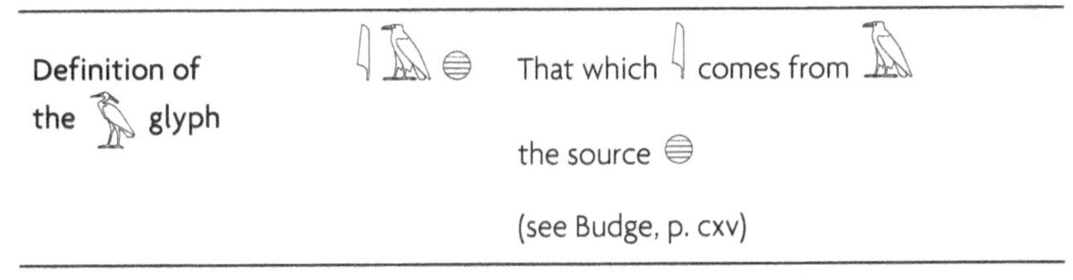

Definition of the glyph		That which comes from the source (see Budge, p. cxv)

Griaule and Dieterlen tell us that, in Dogon culture, the symbol "plays the role of conveyor of knowledge." In fact, they say that "the development of Dogon thought, and hence the elaboration of concepts, proceeds by analogy and has constant recourse to the symbol."[9] Likewise, they say that the common use of symbols in Dogon life serves to familiarize the tribespeople with the concept of abstractions, defined in terms of successive symbols. Griaule and Dieterlen describe the existence among the Dogon of an "immense system of drawn signs expressing the sum of Dogon knowledge." Within this system based on symbols, a myth takes on the role of a kind of elaboration whose first purpose is to broadly explain a topic whose component concepts are defined by symbols. The Dogon term *aduno so*—literally "symbolic word"—refers to a kind of symbolic expression that "simultaneously describes and comprises" the universe.

If we were to summarize the perspective on symbols that seems to be reflected in the various preceding Egyptian hieroglyphic words, we might conclude, in accordance with definitions provided by the Dogon priests, that in its most basic form, a symbol is defined as an action that is intended first to preserve, then to recreate knowledge. The suggestion is that the most effective symbols involve actions that also cause us to associate meanings with an image or shape. An adequate symbol, as defined in Buddhism, is a symbol—much like the Dogon term *aduno so*—whose unique shape and inherent meaning cannot be lost, because according to our understanding of the concept, both are ultimately observable in nature. Set down in this context, the concept of a symbol is expressed as being similar to a mnemonic device, an image or action designed to help us recall an important fact through direct association. Fundamentally, to symbolize is to perform a ritual act, the act of "binding a matter with the matter's image." An image is defined as something that we can see or observe. The simple act of seeing that image keeps us in mind of—and reinforces—the meaning that was intended to be conveyed.

Whether a symbolic act focuses meaning on a drawn character, a sculpted icon, an everyday object, an animal, or an aligned ritual shrine, the basic concept remains unchanged: to symbolize is to perform an act that is intended to convey meaning. The significant emphasis is first on the act or action by which meaning is instilled and recovered, and only secondarily on the specific icon that becomes the object of symbolic meaning. The example of an aligned ritual shrine such as the stupa illustrates this point well. With a stupa, concepts of cosmology are conveyed more through the processes and component shapes that are evoked during its construction than by any aspect of the completed structure itself. Seeing the structure reinforces our knowledge of those shapes and concepts.

This expanded definition of symbolism is wholly consonant with the notion of a symbol as it is practiced among the Dogon. From the Dogon perspective, cosmological symbolism is perceived in common objects and everyday acts of Dogon life. In such a system, key aspects of the cosmology are continuously refreshed and reinforced in the minds of the Dogon tribespeople through their own simple daily acts and through ongoing contacts with everyday symbolic objects that surround them. For example, whenever a member of the tribe heats food in a clay pot, it is only natural that he contemplates a fireside myth that compares the sun to a clay pot raised to a high heat at the time of the creation of the universe. Likewise, when a pale fox passes by—the animal whose tracks are said by Dogon diviners to portend the outcome of future events—the tribesperson naturally tends to think about judgments that must be made to distinguish between truth and error in her own life.

The idea of "binding a matter with the matter's image" is one that speaks to a more visceral aspect of symbolism and suggests a dimension of relational meaning for symbols that can play no comparable role in our understanding of a simple phonetic letter in the modern sense. First, the idea implies that there are specific images that are deemed to best express a given concept. This is the same implication that lies at the heart of the definition of an adequate symbol in Buddhism. From this point of view, it seems only natural that an electron can be symbolized by the shape of its own orbital path as it surrounds the nucleus of an atom. This same intimate association between image and meaning also speaks to the capability of human society to sustain a cultural metaphor over very long periods of time. For example, we might not imagine that, nearly five thousand years later, mankind would still reflexively associate the image of an owl with the concept of knowledge or wisdom, the very same symbolic value we interpreted for the ancient Egyptian owl glyph  in *Sacred Symbols of the Dogon.*

If we can say that symbolism begins with the establishment of an identity between a concept and the attributes of its associated shape—its "image," as the Egyptian word suggests—then it becomes an easy task to see how other images drawn from nature might have come to be adopted to symbolize concepts that may have no inherently associated shape of their own. An example of this kind of adopted symbolism is illustrated by the symbolic assignment of the concept of vibration to the image of a hare or rabbit  . In this case, the identification is between the concept of vibration and a key attribute of the animal represented—the hare's tendency to twitch. The attribute of the animal is, in effect, both highlighted by and affixed to the associated concept. Figuratively speaking, in each case of adopted symbolism of this type, the animal whose image is chosen to symbolize the concept provides us, through its salient attributes, with a working example of the concept that can be observed in nature. Such an association effectively elevates the adopted pairing of concept and image somewhat closer to the level of an adequate symbol.

Whereas the image of an adequate symbol such as an electron is one that, because of its microscopic size, cannot readily be seen without specialized technical equipment, it is also one that, because of the unchanging nature of its inherent shape, cannot ultimately be lost, even through prolonged miscommunication, inaction, or neglect on the part of generations of initiates. Conversely, while the image of an adopted symbol that is drawn from nature can be readily seen, its proper symbolic meaning must either be so very obvious as to effectively constitute an adequate symbol or else must be actively transmitted from initiate to initiate or recovered by later initiates through correct reinterpretation. Snodgrass writes, "According to the traditional Indian concepts of the symbol, meanings are not 'read into' symbols or added to them as a conceptual garnish. On the contrary, they are deemed to inhere within the form of the symbol. . . ."[10] Dogon cosmology provides us with a number of illustrative examples of the different classic forms that a symbolic act can take. These include:

1. **Ritual acts that mimic a process of creation.** For this type of symbolic act, the physical steps by which the act is carried out are defined as replicating a process of creation. Perhaps the quintessential example of this kind of symbolic act is reflected in the building of a Dogon granary or Buddhist stupa. On one hand, a stupa is said to symbolically manifest an "ordered space" from a "disordered field," much as, in astrophysics, an act of perception is thought to evoke the formation of space as we know it from massless waves. From another perspective, the base plan of the granary or stupa begins with the Dogon egg-in-a-ball shape, which is correlated by the Dogon priests to a fertilized egg in a womb, whose growth proceeds through a series of subdivisions in pairs. This stage of construction of a stupa can be seen to replicate the process that a fertilized egg passes through as it divides to become a zygote. The suggestion is that a similar process may also describe the as-yet-undetermined stages that a perceived wave goes through as it attains mass on its journey to form particles of matter.
2. **Ritual shapes that recreate shapes that are observable in science.** This type of ritual shape, referred to as an adequate

symbol in Buddhism, is defined as a symbolic shape that does not lose its cosmological meaning, even if its overt definition is lost over time to the chain of initiates in the esoteric tradition. As we have mentioned, a good example of this type of shape would be the nest drawing in Dogon cosmology, which presents the same image as one of the typical orbital paths of an electron as it surrounds the nucleus of an atom. Other examples would be the highly recognizable shapes of string intersection diagrams in string theory . Such shapes hold direct meaning for modern researchers of cosmological science, so they may be properly understood within that context, even if all trace of esoteric meaning relating to the symbols is lost.

3. **Symbolic concepts that are defined through narrative.** This kind of symbolic definition is often found in the storylines of Dogon myth. These are typically described as events that are supposed to have transpired at the time of the formation of the universe. Through the narrative of the myth, often-restated equivalences are established between important cosmological concepts and corresponding symbols. Examples of such symbols include depictions of the unformed universe as a primordial egg or the concept of a particle rendered by the image of a clay pot.
4. **Mythological characters whose names, actions, or attributes define component stages of creation.** Perhaps the quintessential example of this type of symbol is found in the character of Ogo, who plays the role of light in Dogon cosmology and whose name recalls the Egyptian word *aakhu,* meaning "light." We are told by Griaule and Dieterlen that the word *ogo* in the Dogon language means "quick" and that Ogo's actions are understood by the Dogon priests to occur out-

side of the context of space and time as we know it, so they are deemed to prefigure the formation of space and time. Another good example of this kind of mythical character is the Dogon god Amma, whose dual aspects, taken from the perspective of biological symbolism, correlate well with the male and female "seeds" required to initiate a fertilized egg in biological reproduction.

5. **Recurring ritual acts whose timing reflects key values in creation science.** This type of recurring ritual is exemplified by the Dogon Sigui festival, which is associated with the stars of Sirius and which recurs in practice every fifty years. Fifty years is the approximate orbital period of the sunlike star Sirius A and its binary partner, the dense dwarf star Sirius B. Other examples of this kind of ritual act might include the observance of seven days of shiva in Judaism, which is closely associated with a death, or a circumcision performed on the eighth day after the birth of a child. In Dogon cosmology, the seventh stage of vibration of a primordial thread is characterized as the death of a primordial egg, which correlates to a Calabi-Yau space in string theory, while an eighth stage of piercing, tearing, or cutting is associated with the birth of a new egg.
6. **Various four-level symbolic progressions used to define processes of creation.** In Dogon cosmology, this kind of progression is perhaps most vividly illustrated by the terms *bummo, yala, tonu,* and *toymu,* which are understood as the four successive stages of any creative act. Another more familiar example might be seen in the four primordial elements—water, fire, wind, and earth—which are defined in ancient cosmology as the four primordial component states of matter. These elements are interpreted in *Sacred Symbols of the Dogon* as representing four progressive stages in the formation of mass.

Griaule observed in *Conversations with Ogotemmeli* that, strictly speaking, Dogon cosmology cannot be called an esoteric tradition, since by design its inner details are held open to any tribesperson who cares to systematically pursue them. This outlook is seemingly affirmed in Buddhism by the very definition of an adequate symbol, which offers us the unvarnished image of an important cosmological shape in plainest view, with no attempt at deliberate obfuscation. The ultimate meaning of such a symbol is hidden from us only by our own limited understanding of cosmological science. Thus, it appears that the original nature of an adequate symbol was to convey a scientifically precise image and meaning, as opposed to some obscured or hidden meaning, but only for an individual who is already scientifically aware. So the Dogon cosmological tradition seems designed to transmit such symbols as an intact body through generations of initiates. If the underlying scientific meanings we propose for these symbols are valid, then many of these initiates might never have hoped to acquire a literal understanding of their scientific import; rather, the role of these initiates may simply have been to deliver—whole—the larger body of symbols to some future generation that would be more capable of recognizing and appreciating their most essential meanings.

As suggested in previous volumes, the logic that underlies the assignment of meaning to a single symbol in Dogon and Egyptian cosmology can often be seen to extend to progressions of related symbols. For instance, creation is defined as being from water 𓈖, and matter starts out in wavelike form 𓈗. From this perspective, the processes by which waves are collected together to form larger bodies are sensibly represented by everyday objects that are used to contain water, such as bowls or basins 𓎟 or cups ▽, while particles themselves are represented by clay pots ○. In this way, the daily Dogon act of collecting water comes to be equated with the processes of the formation of particles of matter from primordial waves.

According to Gerald Swanson, author of *Explorations in Early Chinese Cosmology,* the traditional dualism that is a stated principle of Dogon cosmology and culture may also be reflected in the roles of the earliest symbolic references of ancient Chinese cosmology. In his book, Swanson describes the interplay of two pivotal symbolic concepts called *hua* and *shen,* as described in the ancient Chinese *Book of Changes.*[11] The concept of hua was associated with *images,* a Dogon cosmological term meaning "symbols," and with measurements, which Swanson says most often relate to transformations that are similar to the component stages of creation in Dogon cosmology. These are the concepts typically defined by Dogon cosmological drawings and their likely Egyptian glyph counterparts. Shen was a notion linked with rituals such as the Dogon Sigui festival, whose period of recurrence was tagged to the orbital period of the stars of Sirius, and ritual circumcision, which is said by the Dogon priests to represent the circular orbit of the stars of Sirius. According to Swanson, such rituals deal primarily with movements such as the risings, settings, and orbits of astronomical bodies. Similar concepts are reflected in the Egyptian hieroglyphic language by the words *shenn* and *shenu,* both of which refer to the concept of an orbit.[12] The Egyptian word *shen aten* refers to the circuit of the solar disk. An Egyptian word for granary, the Dogon ritual structure whose faces were associated with the risings and settings of stars related to planting and harvesting, is *shna.*[13] Meanwhile, the Egyptian word *uha* refers both to the stage of a journey and to the act of measuring. (Budge defines it as "to stretch out a builder's cord to show the size of a building.") Clearly the Chinese principles of *shen* and *hua,* even if not overtly expressed by the Dogon or Egyptian priests, seem to play out in practice in Dogon and Egyptian cosmology and language.

In *Sacred Symbols of the Dogon,* we demonstrated ways in which the family of symbols that compose Dogon and Egyptian cosmology relate to each other under the umbrella of broad guiding metaphors whose purpose seems to be to help lead us through a given progression of symbols. For example, the generic concept of a force in the scientific sense is conveyed by the figure of a bent arm 𓂝, which can be seen as the primary agent by which force is exerted in daily life. The notion of an effect, such as the slowing of the time frame that accompanies an increase in mass or acceleration, is conveyed by the figure of a bent arm holding an object, such as a twig or blade of grass 𓂡.

Symbolic statements that relate to the three most similar quantum forces—gravity, the weak nuclear force, and the strong nuclear force—are given in terms of human figures: falling 𓀜 (gravity), bowing 𓀠 (the weak nuclear force), and standing firm 𓀗 (the strong nuclear force).[14] The larger set of drawn shapes used to represent the component stages of matter are organized according to a series of grand metaphors that are familiar to human experience, such as the stages of growth of a plant from a seed to a grown plant or the growth of an adult bird from an egg.

FIVE

GUIDING METAPHORS OF THE COSMOLOGY

The more we study various ancient cosmologies, the more apparent it becomes that they were organized around several ongoing, symbolic guiding metaphors that were apparently introduced to help us conceptualize—in the proper orientation and sequence—various progressive stages of creation. Typically, these are presented as four-stage metaphors, and they are exemplified by the Dogon words *bummo, yala, tonu,* and *toymu.*[1] The Dogon priests explain these words in terms of four conceptual stages in the construction of a house, stages that are comparable to those required to produce an architectural drawing. This seems particularly appropriate because the defining act of the cosmology seems to have been the construction of an aligned ritual structure such as a stupa. At the first stage in this construction metaphor (*bummo*), a location has been selected for a structure that only exists in a concept, as the Dogon priests would say, in signs rather than in actuality. At the second stage (*yala*), the rough dimensions of the structure have been defined by stones that have been plotted as endpoints to define the gross outer dimensions of the building. In the third stage (*tonu*), more stones have been placed to "flesh out" key features of the structure such as the location of walls and positions of doorways. In the final stage (*toymu*), the structure is complete. Likely Egyptian correlates to these terms were identified in *The Science of the Dogon,* and the likely pronunciations (the actual pronunciation of Egyptian words is uncertain) and meanings are commensurate with those of the Dogon. These include the Egyptian word *bu maa,*[2] which means "place perceived," the word *ahau,*[3] which means "delimitation posts or boundaries," the word *teni,*[4] which means "to estimate," and the word *temau,*[5] which means "complete."

The Dogon priests say that the concepts represented by these four stages can be applied to any creative act. In fact, the definition of this four-stage progression prepares us, and is perhaps meant to orient us, to conceptualize key aspects of creation as four-stage progressions. As we proceed in our studies of Dogon and Egyptian cosmology, we will see these conceptual stages represented through various symbolic methods, as if to shepherd us through these stages of creation. The first, and perhaps most widely familiar, of these metaphors can be found in the four primordial elements—water, fire, wind, and earth—which are likely examples of four states of matter that the Greek philosophers took symbolically to represent constituent parts of matter. In chapter 7, we will show how these four elements appear diagrammatically in the egg-in-a-ball figure; they emerge conceptually just after the initial perception of a massless wave, so they would most properly correspond to the initial appearance of mass. Together, they represent four component stages in the formation of mass, and each is assigned a symbolic key word. We interpret these to be (1) massless waves (water), (2) an initial act of perception (fire), (3) vibration (wind), and (4) mass (earth).

In much the same way, broad metaphors have been established within the cosmology to represent four stages in the formation of an atom. Again, as illustrated in *Sacred Symbols of the Dogon,* these metaphors are depicted in the Egyptian hieroglyphic language by a series of glyphs, each also associated with appropriate key words of the cosmology. The first metaphor is suggested and introduced by the egg-in-a-ball figure itself; it correlates phases in the growth of matter to stages in the growth of a bird from an egg. These stages, which are illustrated here pictorially in glyph form by images of (1) an egg, (2) a chick, (3) a standing bird, and (4) a flying bird, would be familiar to the experience of Dogon or Egyptian initiates and would provide them with comparative references by which to organize defining symbols of the cosmology.

A second of several such metaphors, seemingly provided in case the initiate should somehow miss the symbolism of the first, involves four similar stages in the growth of a plant from a seed to a grown plant. Again, this symbolic metaphor begins with the egg-in-a-ball's defined relationship to primordial seeds or signs and is closely associated with the cosmological concept of the tree of life. Support for this view is provided by the word *stupa* itself, which can mean "stem" or "trunk." On one level, the stupa is the symbolic form that grows from the initial symbolic seeds, which are themselves defined in terms of the base plan of the stupa. Egyptian words that relate to the concept of the tree of life are typically written using a plant glyph with three branches or shoots, which we take as likely symbolism for the three types of seeds or signs defined in Dogon cosmology: the guide signs, master signs, and world signs. The tree of life itself can be seen as a perfect metaphor for the growth of matter, defined symbolically in terms of the separation of (i.e., creation of distance between) earth (mass) and sky (space), a process that can be equated to the growth of the spreading branches of a tree. Similar symbolism surrounds the concept of the tree of life in the ancient Vedic tradition.

The notion that the Dogon concept of guide signs may have been understood in a similar way in Egyptian cosmology is illustrated by the Egyptian word *up uat,* which means "to open the way, to act as a guide."[6] This word is defined in symbolic terms of the raising of mass to create distance or space, and qualifies under our definition as a defining word for what we interpret as the distance glyph .

TO OPEN THE WAY, TO ACT AS A GUIDE (UP UAT)

Definition of the distance glyph	To raise up or separate mass or matter followed by the distance glyph (see Budge, p. 161b)

Likewise, the Egyptian concept of master signs is also expressed in distinctly Dogon terms. One Egyptian word for master is *ua,* which comes from a phonetic root that means "to be about to do something."[7] This word takes its expression from a cosmological theme that compares the stages of creation to stages in the growth of a plant. It is written alternately as (the raising up or growth of distance) or as (growth comes to be distance). Another Egyptian word for master is pronounced "kherp"[8] and is written , or source of the bending/warping of space.

The Egyptian pronunciation "ua" can be taken as the phonetic root of the important Dogon cosmological word *yurugu,* which is a name given to the pale fox, a central Dogon symbol relating to the creation of matter. According to Dogon myth, it was the revolt of yurugu that introduced disorder into the universe and that initiated the separation of *earth and sky* through the creation of distance.[9]

Indications that other such progressive metaphors may have been intended in the parent cosmology can also be seen when we consider the paired notions of creation from water, which is applied most obviously in cosmological terms, and creation from clay, which immediately brings to mind the reproductive themes of the cosmology. If matter begins in the form of waves, then a particle can be thought of as a kind of collection of waves. In the Egyptian hieroglyphic language, we see images of clay pots —vessels that are used to collect and hold water—appear in association with particles at various component stages of matter. Of course, as we have suggested, clay pots employed as a way to collect water constitute an act of daily life that would be quite familiar to any human society in the early stages of civilization. In fact, one Egyptian word for pottery, pronounced "nehep," seems to express the cosmological concept of pottery in precisely these same terms.

TO FASHION A POT—DEFINITION OF A PARTICLE (NEHEP)

Pottery		"Seeds" in a pot collected together ,
		followed by the held particle glyph
		(see Budge, p. 384b)

As specifically illustrated in *Sacred Symbols of the Dogon* and as mentioned in our earlier discussion of symbolism, a similar approach seems to be taken with regard to the symbolic representation of the four quantum forces—gravity, the electromagnetic force, the weak nuclear force, and the strong nuclear force. In the Egyptian hieroglyphic language, the generic concept of a force is represented as a bent arm , the body part through which force is typically exerted and the same part used to define the ancient cubit unit of measure. The concept of gravity is aptly defined by the picture of a falling man . The concept of the weak nuclear force is assigned to the figure of a bowing man , while that of the strong nuclear force is illustrated by a stocky, upright man .

Another guiding metaphor of the cosmology that has been

mentioned previously is one that is also central to the plan for zoological classification of animals that relates to the Dogon granary. This involves four families of living creatures and progresses from insects to fish to four-legged or domesticated animals to birds—the four major classes of animals associated with the four central staircases of the Dogon granary. By this view, concepts of nonexistence coming into existence would be represented by an insect, in Egyptian cosmology by the dung beetle. Concepts relating to the initial raising up of mass or matter would be defined by the fish, in Dogon cosmology by the *Nummo* fish. Matter in its state of disorder and reorganization would be represented by a four-legged animal, in both Dogon and Egyptian cosmology by the jackal, dog, wolf, or fox. Matter in its completed state would be symbolized by a bird, in the Egyptian concept of completed existence (pau or pau-t) by a flying goose.

Other guiding metaphors are evident throughout Dogon and Egyptian cosmology and language. One of the most obvious of these involves concepts relating to the formation of the spoken Word. These references span the various stages in the creation of the po pilu or the Calabi-Yau space. By this metaphor, the Word is spoken at the eighth stage in which the egg of the world, the po pilu, is pierced.[10] This metaphor has the added effect of orienting us to think of matter, like the act of speech, as being a fundamental product of vibrations. Another similar metaphor is found in the concept of weaving, a notion that pervades the cosmology and that orients us to think of matter as also being, at root, a product of primordial threads.

As a final point, we might note that the ease with which Egyptian cosmological words can be shown to consistently relate to appropriate images of the cosmology offers testimony to the idea that early Egyptian glyphs may well have taken their shapes and meanings from a preexisting cosmology.

SIX

THE EGG-IN-A-BALL

One important but rarely discussed aspect of Dogon cosmology is that Amma, the creator god, is not large and powerful, but instead so very small as to be effectively hidden from view.[1] If, in truth, one purpose of Dogon cosmology is to define how matter is formed, then the true nature of Amma is pivotal to an understanding of the innermost purposes of the cosmology, so it becomes an esoteric detail that, in the view of the Dogon priests, is not to be discussed in public. Consequently, the first among many important drawings of Dogon cosmology that we should discuss is the Picture of Amma, which is alternately known as the Womb of All World Signs or simply as the egg-in-a-ball.[2] This figure represents the conceptual starting point of creation from both a cosmological and a biological perspective, the point at which the transformation of matter (from its wavelike state into particles) and the conception of life (from a fertilized egg to an embryo) begins. In my opinion, this figure, which we will see ultimately bears an important relationship to the Dogon po pilu, or egg of the world, and to the Calabi-Yau space of string theory, represents the starting point, or conceptual bottom, of the cosmology.

From the perspective of the creation of matter, creation rests for the Dogon on 266 fundamental seeds or signs, mythical components of matter that likely correspond to the more than two hundred fundamental particles known to modern astrophysicists. On the most superficial level, the Picture of Amma is meant to help us conceptualize how the Dogon organize these signs. As we have suggested, these particles are divided by the Dogon priests into three distinct classes: two guide signs, eight master signs, and 256 world signs. On one level, the figure of the egg-in-a-ball is meant to illustrate (in two dimensions) the evolution of the guide signs and the master signs from Amma.

The figure of the egg-in-a-ball begins with what is essentially the shape of the Egyptian sun glyph ☉. This figure consists of a larger outer circle (the ball) drawn around a smaller inner circle (the egg). From the biological perspective of the cosmology, it is clear what the figures represent. The inner figure, which is actually called an egg, represents a fertilized egg or ovum. In case there should be any doubt—notwithstanding the explicit identification of the drawing in its alternate title as a womb—that we are talking about components of a symbolic uterus, this symbolism is underscored by the assignment of the term umbilicus munde, or umbilicus of the world, to the egg. The Dogon priests bisect this sun glyph figure both vertically and horizontally with two intersecting lines, which are said to comprise the axis munde, or axis of the world, thus reinforcing the parallel cosmological symbolism of the drawing. These lines effectively divide the figure into four quarters of equal size that are meant to symbolize the four classic elements of water, fire, wind, and earth—the four primordial components of matter in Greek cosmology. Thus this figure that conceptually initiates Dogon cosmology also can be seen to initiate the two major parallel themes of the cosmology, the creation of matter and the creation of life through the processes of biological reproduction.

In accordance with the figure's title Picture of Amma, we are told that Amma resides symbolically within the central egg. Since in

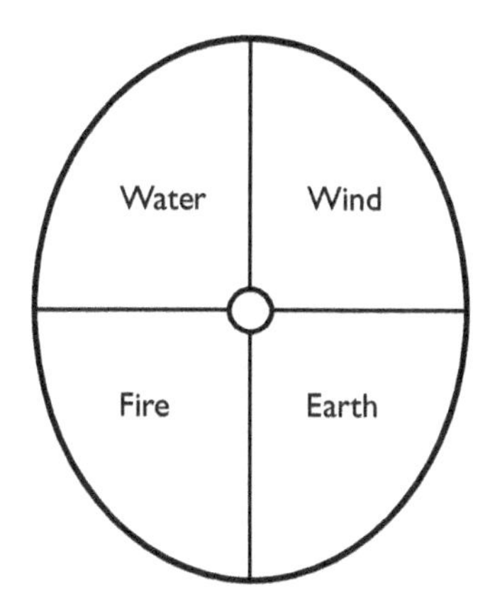

Dogon Womb of All World signs, representation of Amma.

Dogon cosmology there is a principle of duality in the universe, even the creator/god Amma is conceived of as having a dual nature. We can shed potential light on these dual aspects of Amma by looking to the Egyptian words *am* and *maa,* cosmological words that we take as the likely prefix and suffix of the name. The first of these words, *am,* refers symbolically to the concept of knowledge,[3] which in the biblical sense refers to biological conception. The second, *maa,* a likely phonetic root of the important Egyptian concept of *maat* (which is sometimes translated as "truth" or "order"), means "to examine or perceive,"[4] the processes by which truth is discovered. These two concepts, which together form the name of the Dogon god, can also be understood to represent the opposing aspects of creation, consistent with the cosmological and biological reproductive themes of the cosmology. The first, conception, instigates the formation of a fertilized egg, and the second, perception, precipitates the transformation of matter from its wavelike form into what we ultimately perceive as particles. Consistent with this notion of the dual aspects of the creator/god, the Dogon priests tell us that Amma also symbolically represents the two initial guide signs or eye signs of Dogon cosmology.

It is significant that Amma himself is compared to or equated to seeds or grains. The Dogon say that Amma is like the *yu* grain,[5] which takes a shape similar to () and can be seen as a symbolic counterpart to the central egg of the egg-in-a-ball figure. By comparison, one ancient Egyptian word for grain is *amm*,[6] which can be interpreted as a phonetic root for the Egyptian word *amma* or *ami*, which according to Budge means "give, let, grant, I pray, make, cause." Likewise, the Dogon concept of seeds or signs may be reflected in the Egyptian word *tcharm*, which can refer either to a seed or the act of making a sign.[7] (It is interesting that the similar English word *charm* can still refer to a magical amulet or sign in modern usage.) The Egyptian word *tcharm* is written:

TO MAKE A SIGN (TCHARM)

Concept of a sign		Perception , the foremost part of knowledge , followed by the spoken word or sign glyph (see Budge, p. 899b)

Once it has been divided by the axes, each quarter of the Dogon egg-in-a-ball, along with its assigned element, is associated with a pair of the eight master signs. Like the two guide signs, the two master signs associated with each pair represent opposing aspects of their corresponding elements, as outlined in the table below:

ELEMENT	OPPOSING ASPECTS
Water	Wetness versus dryness
Fire	Light versus darkness or fire versus wood (wood is what fire burns)
Wind	Air in motion versus still air
Earth	Earth versus empyreal sky, mass versus space, essence versus substance

Each of the eight master signs is also explicitly associated with one of the eight Dogon ancestors. These ancestors are defined by the Dogon priests as quasi-mythical/historical intermediaries who were temporarily removed from mankind to be instructed in the civilizing plan of the cosmology, then delegated to transmit that plan to the rest of humanity. The Dogon priests say that the authors of the cosmology chose to use intermediaries because they were fearful of what the effect might be of prolonged contact between purely spiritual beings such as themselves and creatures of flesh and blood. (This view of Dogon cosmology is substantially upheld in Buddhism, where, according to Snodgrass, the "most sacred" of the Buddhist symbols were deemed to have been given to mankind by a "non-human source.")

In the Dogon view, all things in the universe are "manifested by thought."[8] One Egyptian word for thought, *amiu-khat*[9] is based on the phonetic root *am*. It is also very similar to the Egyptian words for ancestors, *amiu-hat,* and for followers or those who come after, which is *amiu-khet.*

It would be difficult for an observant comparative cosmologist not to notice that the figure of the egg-in-a-ball defines a symbolic construct or framework within which the creator god and the eight paired deities of the Egyptian Ennead can be neatly understood. Much like the Dogon ancestors, the cosmological principles of duality and the pairing of opposites are upheld by the very similar manner in which the Egyptian deities emerge in male/female pairs. To the extent that the Egyptian symbolism is clearly understood, it also upholds the notion that these ancestors/gods might have borne an original relationship to the four primordial elements, as is illustrated with the traditional symbolism of the earth god Geb and the sky goddess Nut or the Egyptian god of the air Shu. As is acknowledged by traditional Egyptologists, the pantheon of Egyptian deities, along with their associated family tree, has shifted over time, so we are often left to infer what might have been true about the cosmology in the earliest days of Egyptian culture.

The outward similarities that we have noted between the shape of the egg-in-a-ball and the Egyptian sun glyph ☉ will be made even more evident in regard to the shape's relationship to the concept of the aligned ritual shrine, as we will discuss in the next chapter. Careful study of the cosmology makes it clear that this same shape is explicitly associated with the sun, both by the Dogon priests and the Buddhists, as well as other early cultures such as the ancient Chinese. Since the figure is interpreted by the Dogon as a picture of the creator/god Amma, then the traditional symbolic assignments that were made in many ancient cultures between the sun and a creator/god become more understandable, and their likely purpose comes more clearly into focus.

Taken together, the figure of the egg-in-a-ball is meant to represent the first conceptual stages in which existence emerges from non-existence, and by which order comes to be manifested from disorder. As we mentioned previously, it is clear that in the original cosmology, symbolic meanings were not casually assigned to their associated images. Rather, we have seen that the key attributes of an image or object in these ancient cosmologies most often make it a truly compelling choice to convey the meaning of its assigned symbolic concept. We can illustrate this point by considering that, from the point of view of the structure of matter, modern speculation regarding the Calabi-Yau space (the likely scientific counterpart of the Dogon po pilu, or egg of the world[10]) begins with a construct called the E8 figure, a mathematical conception of the wrapped-up dimensions of string theory or torsion theory. This figure, when represented in approximation in two dimensions, takes a form that is distinctly reminiscent of the Dogon egg-in-a-ball shape.

Similarly, from the perspective of biological reproduction,

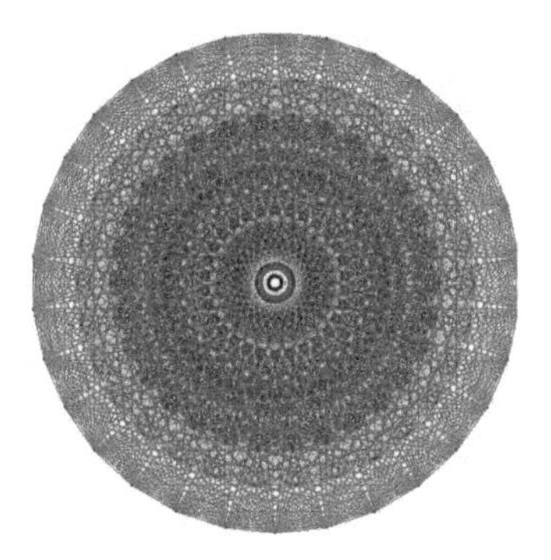

Diagram of the E8 shape in two dimensions. From http://aimath.org/E8/mcmullen.html.

scientists agree that the creation of life begins with the DNA molecule. In 1953, British biophysicist Rosalind Franklin produced an X-ray diffraction image of a DNA molecule as it actually appears when viewed from the top down. This image is a readily recognizable counterpart to the Dogon egg-in-a-ball figure, complete with segmenting lines reminiscent of the axis munde that divide it into roughly equal quarters.

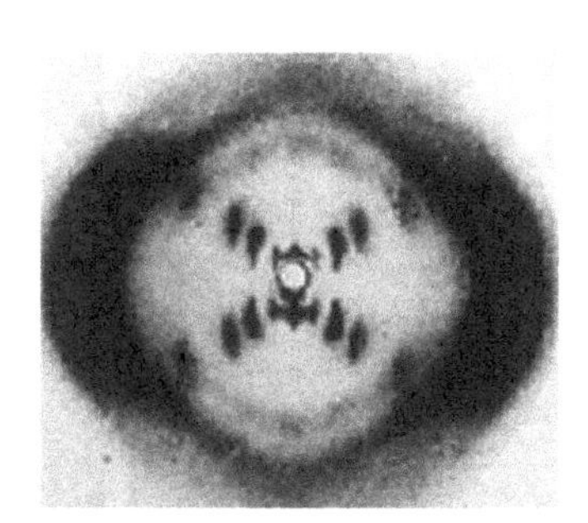

X-ray diffraction image of a DNA molecule. From http://scienceblogs.com/bioephemera/2008/04/juxtaposition_5.php.

In each case, and in terms of both major themes of the cosmology, there is ample evidence to suggest that both the figure and symbolism of the Dogon egg-in-a-ball is in substantial agreement with concepts and images as they are understood in modern science, so it constitutes a wholly appropriate symbol to represent those concepts.

As a symbol, the egg-in-a-ball can be interpreted in a variety of different ways in ancient cosmology. For example, the traditional Chinese symbolism that relates to the concepts of yin and yang can be understood to fall neatly in line with Dogon concepts that relate to Amma. The traditional Chinese figure that is used to depict the relationship between yin and yang is that of a circular egg—like the Dogon egg—composed of two opposing yet complementary aspects, not unlike the Dogon definition of the two guide signs that are symbolized by Amma.[11] The yin/yang figure even includes two opposing smaller circles, the same figure as the Egyptian glyph that is traditionally interpreted to represent the concept of seeds or signs.

The Chinese yin/yang figure

SEVEN

THE ALIGNED RITUAL SHRINE

At first glance, even without a broader understanding of the deep principles that may underlie ancient cosmology, the aligned ritual shrine can be seen by an observant student of cosmology as one of its most pervasive symbols. Such structures, which are found in widespread regions of the world in conjunction with other signature elements of cosmology, find practical expression in many widely varying forms, ranging from cairns in the British Isles, to ritual mounds found in North America and elsewhere, to ancient henges in the British Isles, to astronomic circles found worldwide, to the mastaba in ancient Egypt, and ultimately to the pyramid, perhaps the quintessential ancient aligned ritual structure, which is also found worldwide.

From the perspective of the Dogon priests, it is clear that, on the most fundamental level, the aligned ritual shrine can be interpreted as a three-dimensional representation of the egg-in-a-ball drawing—the Picture of Amma—which is meant to mnemonically recreate concepts of the cosmology. From this viewpoint, it carries with it all of the symbolism that has been previously described in relation to that figure. As a symbol of the act that is understood to have catalyzed creation, the construction of a granary or stupa can be seen as a real-world reenactment of the manifestation of creation.[1]

Snodgrass states in his book *The Symbolism of the Stupa* that, like the idealized form of the Dogon granary, which represents a conceptual tool, the stupa is conceived in Buddhism as a purely symbolic form, one that, by intentional design, is expected to serve no functional purpose that might detract from its primary role as the master symbol of a system of cosmology.[2] Conversely, in the desert culture of the cliff-dwelling Dogon, where subsistence itself can often be a challenge, it can be argued that every structure must serve a productive purpose, so Dogon granaries, which, like modern-day stupas, are most often not built according to the classic cosmological model, are used to store grain. However, from a priestly perspective, the ritual granary of Dogon cosmology, like the Buddhist stupa, is treated as a classic symbol, more as an idealized concept than as a practical architectural form.

For the serious initiate, both the Dogon and Buddhist cosmological systems begin with the aligned ritual shrine. Consequently, the structural and symbolic attributes of the shrine also represent the most sensible starting point for comparisons of Dogon, Buddhist, and Egyptian cosmologies. Cross-confirmation of statements that are affirmed within each of these cosmologies with regard to the form and symbolism of the ritual shrine provide a factual basis for these comparisons. Close examination of two ancient Egyptian hieroglyphic words for shrine reveals a conceptual mind-set reflected within the glyph structure that is in substantial agreement with the Buddhist outlook on the symbolic nature of the shrine as outlined by Professor Snodgrass.

The first of these is the word *skhem,* which is written . Symbolically, the word begins with the glyphs , which spell the Egyptian word *sekh,* meaning "to grasp,"[3] an important concept we've previously discussed in relation to the creator gods Amma and Amen, and about which we will have more to say in chapter 11. Symbolically, we read these two glyphs as meaning "to bind with the source." The owl glyph is a character that we interpret in its traditional symbolic sense to mean "knowledge." The glyph, in the view of a traditional Egyptologist, implies the concept of negation.[4] The final glyph is a figure that we associated in *Sacred Symbols of the Dogon* with the concept of a structure.[5] The first four of these glyphs in the written word *skhem*, taken together, spell the Egyptian word *s-khemi*, which Budge defines as meaning "to be unmindful of, to forget."[6] Taken together, we interpret these glyphs symbolically to read:

TO BE UNMINDFUL OF (S-KHEMI)

To be unmindful of		To grasp knowledge
		not
		(see Budge, p. 616a)

Based on the symbolic reading above, we can substitute the meaning "to be unmindful of" in place of the leading glyphs of the word *skhem* and derive a meaning that is reflective of the Buddhist concept that a shrine is intended to be a symbolic, rather than a strictly functional structure:

SHRINE (SKHEM)

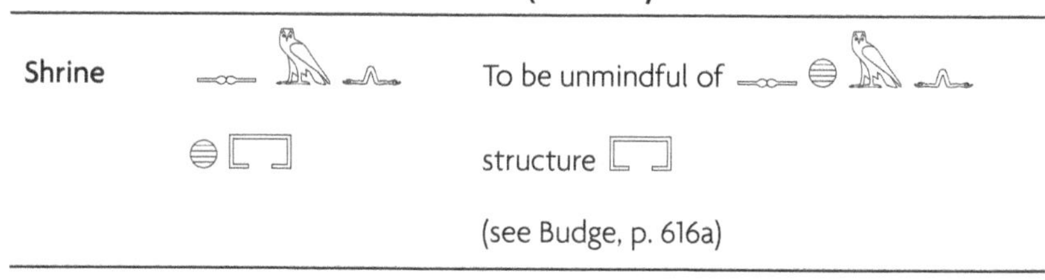

Shrine		To be unmindful of
		structure
		(see Budge, p. 616a)

The Dogon granary, like many other symbols of Dogon life, blends what is understood to be an important symbolic form with a functionally useful structure. Although they seem to have gone unnoticed for the half-century since Griaule and Dieterlen first studied the Dogon, the similarities between the Dogon granary and Buddhist stupa as aligned ritual symbolic structures run deep. Both serve as the conceptual starting point for their respective, parallel cosmologies. Both evoke creational symbolism that is interpreted symbolically on two levels at once: one that is cosmological, the other biological in nature. Both define the same sequence of evoked ritual shapes. Both conjure matching sets of cosmological concepts and symbolic meanings associated with those shapes. Both play the same pivotal role in early creation myths of their respective cultures that describe the deliberate, informed instruction of civilizing skills to humanity. Both are interpreted on one mythical level as representing the actual vehicle on or within which a revered mythical ancestor/teacher descended from the heavens to impart civilizing knowledge.

At heart, it is the parallel attributes of these ritual aligned structures—the Buddhist stupa and the Dogon granary—along with the extensive symbolic systems that they evoke, that allow us to effectively synchronize the Dogon and Buddhist cosmologies. It is the common aspects of these systems that provide us with a working framework within which to interpret many otherwise obscure references of ancient Egyptian cosmology that, as we have shown, often present themselves in similar forms. Likewise, it is by way of such parallels that Buddhism can be seen to independently affirm many of the esoteric details of Dogon cosmology as reported by Griaule, which affirms it as a legitimate and wholly sensible system of cosmology.

The Buddhist and Dogon cosmological traditions are in agreement about several key functions of the aligned ritual shrine. First, both tell us that the stages in construction were meant to mimic important processes of biological and cosmological creation. This means that, when we observe how a stupa is created, we gain potential insights—not always through literal representation, but more often through analogy—into how biological reproduction and the manifestation of matter may occur. Second, both traditions agree that the aligned ritual structure was intended to serve as a mnemonic for the cosmology itself, which was considered in both cultures to define a world system or world plan. The idea is any initiate who could successfully build a Buddhist stupa or Dogon granary would, through that very process, at the same time recreate many of the key conceptual elements of the cosmology.[7] On another level, the building of a stupa or granary also serves as a practical exercise in skills that are prerequisites to civilized life, skills that relate to the planning, measuring, plotting, and building of a simple structure.

To understand fully how this process may have progressed, we must first integrate complementary Dogon and Buddhist statements regarding various aspects of the stupa and granary plans to produce an integrated view of how their construction may have been intended to play out. According to Snodgrass, an initiate who intends to build a stupa or granary begins simply with a stick and an uncultivated field, two resources that were certain to have been plentiful among the nomadic hunter tribes of predynastic Egypt. The initiate must:

1. Place the stick vertically in the ground and draw a circle around it with a radius twice the length of the stick. (In regard to the Dogon granary, both the vertical measurement and the radius of the circle are defined as ten cubits. This suggests either that the cubit unit of measurement was established prior to the introduction of the granary form, which would make sense, since a unit of measurement would have been required, or that Dogon culture preserves an Egyptian tradition that dates from

sometime after the establishment of the cubit as a unit of measurement.) The combination of the stick and the circle, whose image ⊙ recalls the image of the Egyptian sun glyph, constitutes an effective sundial that, with minimal effort, the initiate can use to track the hours of the day and to measure a day's length. The circular bases of both the stupa and the Dogon granary are explicitly said in their respective cultures to symbolize the sun. Snodgrass says that the alignment of this figure is determined by reference to the apparent movements of the sun.[8] It becomes clear through discussion regarding both cosmologies that, on another symbolic level, this circle also represents the initial expansion of space within the cosmos at the time of creation or a womb after the time of conception.

2. Mark the two longest shadows of the day—morning and evening—at the points where they intersect the circle.
3. Draw a line between the two marked points. This line will be oriented automatically along an east-west axis and will pass through the stick on two specific days of the year—the two equinoxes. On all other days, the line will move progressively farther away from the stick until the date of the next solstice. Observation of this line and its movements relative to the stick allows the initiate to track the movements of the sun over more extended periods of time, to observe and predict the dates of the equinoxes and solstices, and to calculate the length of a year.
4. Use the two points of intersection as the centerpoints for two additional circles, each drawn with a radius slightly larger than the original circle. These two new circles will intersect each other at two new points.
5. Draw a second line between these two new points of intersection. This line will be oriented automatically along a north-

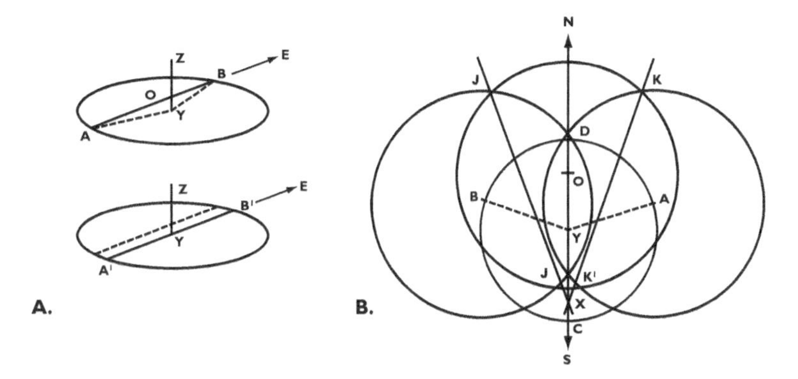

The determination of the orient. From Snodgrass,
The Symbolism of the Stupa, *15.*

south axis and will intersect the first line at a right angle. On the equinox, the lines will bisect the circle and so divide it into four equal parts.

In accordance with the two levels of symbolism associated with the cosmology—biological and cosmological—the Buddhist and Dogon traditions can be seen to be in substantial agreement with each other and with the Dogon egg-in-a-ball drawing, in that each considers these two intersecting lines to represent both the axis munde (axis of the world) and the umbilicus munde (umbilicus of the world), and to define the cardinal directions of north, south, east, and west. The center of the axis is considered to be the *omphalos* or navel.[9] Consequently, the original circle of the base plan can also be seen to represent a womb, and the "egg" within the outer 'ball' of the womb to represent a fertilized egg.

Both traditions hold that matter is the product of woven threads. The two intersecting lines that each bisect the circle create the ✕ shape of the first of three types of string intersection in string theory, the very shapes in science that could be reasonably said to depict the weaving of matter.

The central looped figure that is created by the overlapping intersection of these two new circles provides us with an example of the second type of string intersection in string theory. In Buddhism, this figure is referred to as the fish, and it suggests an underlying concept similar to that of the Nummo fish in Dogon cosmology, which we will discuss in more detail in chapter 12.

6. Let the vertical stick define a third axis, perpendicular to the first two. Together, these three axes represent the deployment of space in six directions from a cosmic center. The six rays, in combination with the center (which, according to Snodgrass, constitutes a seventh ray) define creation, by analogy, in terms of seven sunlike rays. This is a signature aspect of both the po pilu, or egg of the world, in Dogon cosmology and the Calabi-Yau space in string theory or torsion theory. In Buddhism, the vertical ray defines a third axis that is conceptualized as the trunk of the tree of life. (Snodgrass tells us that the word *stupa* can also mean "trunk or stem." According to Budge, the Egyptian word for tree of life is pronounced "aam" or "aama," so it could well relate to the Picture of Amma.) This third axis allows the initiate to define a three-dimensional hemisphere, a figure that is the natural extension of the original circle that forms the base of the stupa or granary. In Buddhism, this hemisphere is defined as symbolizing the concepts of essence or substance, while in regard to the mythological structure of matter defined by the Dogon priests, it corresponds to the concepts of mass or matter. In both cultures, it is equated with the cosmological term *earth,* which we take as a key word for mass.
7. For the Dogon, the granary structure, which begins with a round base, rises to a square, flat roof. If we consider that the

> circular base of the granary represents an initial expansion of space, then it becomes apparent through discussion that the square, flat roof of the structure similarly represents an initial deployment of space.[10] This is a concept that is symbolized in Dogon cosmology by the term *empyreal sky.*

Implicit in the stages of construction of the stupa or granary is the presumption of a working familiarity on the part of the initiate with basic principles of geometry. This includes knowledge of what a circle is and how to draw one, the ability to plot a point, and the ability to extend a line between two points.[11] Likewise, the Dogon priests state that they consider the granary form to define examples of several important geometric shapes, including a circle, a square, and a hemisphere. As for the stupa, rather than having a circular base and rising to a square, flat top like the Dogon granary, the base of a stupa is instead squared, much as the base of an Egyptian pyramid is. The Dogon granary also includes four ten-step staircases, one up the center of each side of its four flat faces, a feature of aligned ritual shrines that is familiar to us in various forms from ancient pyramids found in the Americas.

As previously noted, the dimensions of the granary are given in terms of the cubit,[12] perhaps the quintessential unit of measure of ancient cosmology, and it also provides working examples of key values commonly required in geometry, such as the square of a number, the radius, diameter, and circumference of a circle, and even an approximated value of pi. In Dogon cosmology, the rise and tread of each stair is defined as measuring one cubit, the radius of the base of the granary is defined as ten cubits long, and each side of the square, flat roof is measured as eight cubits. Brief consideration of these numbers reveals a method to their assignment; the area of the roof, which is sixty-four square cubits, is equal to the circumference of the base, if we assign a rounded value of 3.2 to pi.[13] The formula for the circumference of the granary would be:

> Circumference = 2 times pi times the radius of the circle, or 64 cubits = 2 times 3.2 times 10 cubits

In both Buddhism and Dogon culture, the stupa or granary form is symbolic of a world plan or world system,[14] a system by which initiates in the tradition learn to conceptualize and organize the world around them. From this perspective, the steps of the staircases of the Dogon granary symbolize the many zoological classes of animal life, including a wide variety of insects, wild animals, domesticated animals, and birds. Conceptually, each class of animal "stands" on a step of the granary, along with similar animals of other related species. (When questioned by Griaule as to how so many animals could fit on a one-cubit step, Ogotemmeli replied that Griaule should consider these to be purely symbolic steps and suggested that any number of animals of any size can be said to stand on a symbolic step.[15]) Likewise, the interior of the Dogon granary, which includes eight chambers, symbolizes the eight types of grain cultivated by the Dogon, grains that were said to have been brought to humanity by the eight Dogon ancestors/teachers.[16]

Like the stupa, the Dogon granary is considered to be an idealized symbolic form, one that might not always accommodate in actual practice the literal objects that are symbolically attached to it for instructional purposes. In fact, as mentioned in *The Science of the Dogon,* one would face immediate difficulties in trying to apply the dimensions of the Dogon granary to a structure precisely as given. In relation to a base that is twenty cubits in diameter, four ten-step staircases with steps one-cubit deep and one-cubit high would meet at a peak and would not produce the required flat roof eight cubits square. According to Griaule, these kinds of discrepancies did not appear to bother his Dogon informants, who often appeared to value instructional symbolism more than absolute rational self-consistency.

The conceptual underpinnings and ritual justifications for various symbolic assignments associated with the stupa are discussed in great detail by Snodgrass in *The Symbolism of the Stupa,* as are other important symbolic attributes of the granary structure as it is understood within Dogon cosmology by Griaule and Dieterlen. From the perspective of this study, we are primarily concerned with stupa and granary symbolism that would relate to the cosmology and to the instructed civilization of humanity, so the thrust of our discussion here will focus on these aspects of the complex structures. However, simply in terms of the symbolic references mentioned so far, it is clear that the stupa/granary structure was meant to play an important role in the following actions:

1. The definition of concepts of time, timekeeping, establishment of spacial direction, and geographic orientation and measurement
2. The instruction and practical application of construction skills
3. Facilitating and encouraging the practice of careful astronomic observation and record keeping, and establishing a working calendar as support for the future practice of agriculture
4. Providing a working observatory for astronomic observation, along with a mnemonic reference tool for tracking the cyclical stages of the agricultural cycle
5. The introduction of a creation tradition rooted firmly in real science that could be the foundation for ongoing civilization
6. The presentation of an organizational system by which

humanity could conceptualize and categorize important elements of the natural world

On yet another level, the plan of the Dogon granary is designed to reflect the metaphoric stages of cosmological and biological creation. It is from this perspective that we most easily see that the base of the granary recreates the Dogon egg-in-a-ball shape. Conceptually, the granary houses the pairs of guide signs and master signs of Dogon cosmology, symbolized both by the grains stored in its eight chambers and by two grains that are traditionally placed in a tiny cup, located at the point of intersection of two partitions at the very center of the granary.

The likelihood of an underlying relationship between a ritual shrine and instructional knowledge or wisdom is one that is suggested again by the form of the Egyptian word *sa,* which means "shrine."[17] This word is written with the single glyph , which Budge defines in an alternate word entry to mean "back, hinder parts," or when used in conjunction with the owl glyph , "in the following of." Based on our approach to Egyptian language, we would interpret the two glyphs symbolically to read as "in the following of (or in pursuit of) knowledge." This interpretation is supported by an alternate spelling of the word *sa,* which reads:

IN THE FOLLOWING OF/TO GO IN PURSUIT OF KNOWLEDGE (SA)

To go in pursuit of knowledge		Knowledge in pursuit of (immediately behind) to go (see Budge, p. 633b)

A similar interpretation applies to the Egyptian word *sau,* which according to Budge refers to a "sage, wise man, one who is educated." Our interpretation of this word depends again on a glyph , whose likely meaning of "to speak" we will discuss at length in chapter 11. Symbolically, the word reads:

SAGE/WISE MAN/ONE WHO IS EDUCATED (SAU)

Sage or wise man		In the following of (immediately behind) one who speaks . (see Budge, p. 634a)

EIGHT

THE ELEMENTAL DEITIES

The concept of elemental deities—the eight primary ancestors/gods typically defined in many ancient cosmologies and the likely counterparts to the gods and goddesses of the Egyptian Ennead or Ogdoad—is one that may give the appearance of being less elaborated among the Dogon than in Buddhism or in ancient Egyptian cosmology. However, events relating to the emergence of these deities in ancient Egyptian mythology often occur in direct parallel to those in Dogon mythology and suggest that we can trace their likely development to various Dogon forms, beginning with the eight ancestors of Dogon myth. Based on comparisons to the Dogon, we will see again that the concept of a deity seems to be intimately related to the egg-in-a-ball shape and to the aligned ritual shrine in the form of the Dogon granary or the Buddhist stupa.

One observation that we might immediately make is that the ancestors of the Dogon seem to be defined in far less personalized terms than their likely Egyptian counterparts. For example, in Egypt these deities have evolved in any given cult center with well-differentiated identities, explicit lineages, and a well-ordered sequence of emergence in relation to one another. The Ennead gods, as they were known in Heliopolis, traditionally consisted of Ra or Atum, who created Shu (air) and Tefnut (moisture), their children, Geb (earth) and Nut (sky), and their progeny, Osiris, Isis, Seth, and Nephthys.[1] In some cases, other Egyptian gods, such as Horus and Thoth, were added to these, which sometimes resulted in groups totaling more than nine gods and goddesses. Atum, whose initial failed attempt at conception is sometimes compared in Egyptian mythology to a masturbatory act,[2] plays a role that is comparable to that of Amma in Dogon cosmology, whose initial attempt to fertilize the earth is also described as incestuous or masturbatory. These comparisons support a view of Dogon cosmology as a likely earlier or alternate form of the Egyptian cosmology, in which nondeified ancestors are often treated more like dispassionate symbols than as individualized mythical characters or deities.

Likewise, although the specific pantheon of gods that was favored in ancient Egypt has changed over time and within local cult centers, there is a wealth of evidence to suggest a correlation between the Dogon dual-natured god Amma and the Egyptian hidden god Amen. To begin with, the words *amma* and *amen* (or phonetic variants thereof, such as *amon* or *amun*) are equated in the languages of a number of North African tribes, including the Mande and the Bantu. Likewise, the Yoruba, whose cosmology shares many common attributes with that of the Dogon, are known to have worshipped the god Amma, or Amon, whose name for the Yoruba also means "concealed."[3] An ancient goddess Amon, who is known to have been worshipped among the proto-Saharans, is depicted in engravings dated to around 4000 BC with a solar disk and wearing *uraei,* the traditional headdress serpents often associated with Egyptian pharaohs.[4]

In accordance with all of this, it is important to note that, even in the view of Griaule and Dieterlen, none of the named characters of Dogon cosmology appears to rise to the level of an actual deity as the term would be applied in ancient Egypt, except the Dogon hidden god Amma. Consequently, our likeliest approach to explaining the concept of an elemental deity as the Dogon priests understand it would be to carefully consider both Amma, who is explicitly defined by Griaule and Dieterlen as the one true god of the Dogon, and the eight ancestors he symbolically evokes. One primary focus of Dogon cosmology is to define the attributes and actions of Amma, who was instrumental in catalyzing the formation of the universe and matter. As we have mentioned previously, our first impulse might be to imagine that such an important deity must being very large and powerful; however, the Dogon priests tell us that, in truth, Amma is really only very small. We are told that Amma is dual in nature, but that this duality can be interpreted in more than one way. In some cases, Amma is described as having both male and female aspects; in others, as we have suggested, the duality of Amma might be more comparable to the classic concept of opposing forces in some Eastern cosmologies, such as those of yin and yang.

Amma is housed in the central egg of the egg-in-a-ball figure, and in this form is alternately conceptualized by the Dogon priests as consisting of four clavicles,[5] a reference to the four arcs that compose the central egg. These can be interpreted as precursors of the four elements—water, fire, wind, and earth. The Dogon priests describe these arcs as appearing "as if welded together."[6] Viewed in this way, we can make sense of the Dogon belief that all the essential constituents of future space come together as one in Amma. These clavicles border on the four larger sections into which the egg-in-a-ball figure is divided, which are said to house those same four primordial elements and for which the elemental deities are appropriately named.

By our earlier interpretation, the name Amma is initially defined in terms of the Egyptian prefix *am,* which means "knowledge." By way of comparison, one of Budge's Egyptian word entries meaning "egg" is pronounced "amm," and in our view of the Egyptian hieroglyphic language, takes the form of a defining word for the egg glyph ⬯. A symbolic reading of this word seems to explain the Egyptian concept of maa, or perception, in terms that would apply appropriately both to Amma and to the ritual symbolic structure:

DEFINITION OF THE EGG GLYPH (AMM)		
Egg		The act of understanding knowledge , followed by the *egg glyph* (see Budge, p. 121b)

Amma is conceived of as the initial guide signs or eye signs, the first two of the 266 primordial seeds or signs of creation. The Dogon priests tell us that the master signs that populate the outer ball of the Dogon egg-in-a-ball similarly represent the eight ancestors of Dogon mythology. As we have said, these ancestors correlate both in number and attributes to the eight paired Ogdoad or Ennead deities as they are found in ancient Egypt, who are also defined as having emerged as paired opposites.

Budge specifically defines the Egyptian god Amen in terms of these eight elemental deities and calls him "the grandfather of the eight deities."[7] This description mirrors the relationship of Amma to the eight Dogon ancestors as represented by the egg-in-a-ball figure; because of the intervening primordial elements, Amma and the ancestors are conceived of as being two ancestral stages removed from one another.

The name Amen is written , which symbolically means "that which weaves waves," while that of Amen-t, a female counterpart to or aspect of Amen, is written , which means "that which weaves waves of matter" or "that which weaves waves in a womb." These two symbolic readings define the name Amen in the specific terms of the two dominant themes of Dogon cosmology. Another spelling of the name Amen-t uses the very same four leading characters to define the egg glyph . Suggestions that the Egyptian god Amen may have been originally treated more like a Dogon symbol than like an Egyptian deity are reflected in yet another Egyptian spelling of the name of Amen, , which we read symbolically to mean "that which weaves waves (as a symbol)."[8]

We can affirm a likely relationship between the eight elemental deities and the notion of a ritual shrine by examining yet another Egyptian word for shrine, this one pronounced "khem." We interpret this word symbolically in terms that are distinctly similar to previous Egyptian shrine words that we have examined. The suggestion is that an Egyptian shrine, like the Buddhist stupa, is understood to primarily define a symbolic source of knowledge, not a functional structure.

SHRINE OR SANCTUARY (KHEM)

Shrine or sanctuary		Source of knowledge , not structure (see Budge, p. 546b)

The word *khem* carries additional significance for us within Egyptian culture, first because Budge defines it as the name of the "god of procreative and generative power,"[9] the very same two creative processes that have been explicitly assigned to the ritual shrine in Dogon and Buddhist cosmology. Likewise, *Khem* was an ancient name by which Egypt itself was known. But perhaps most significantly, the word *khem* forms the phonetic root of the title *khemenu,* which Budge assigns to the "eight elemental deities of the company of Thoth."[10] When Budge relates Amen to the eight deities, he adds the note, "see *Khemenu.*"[11] Thus, the likely Egyptian counterparts to the eight ancestors defined by the Dogon granary are themselves defined, through their very Egyptian title, in terms of the concept of a shrine. Moreover, through Budge's definition of these same eight gods and goddesses, including Amen, as "elemental deities," he also correlates them definitively with the four primordial elements.

With the key Dogon and Egyptian elemental deities synchronized in this way, we may now begin to correlate various Egyptian gods and goddesses with likely Dogon counterparts, based first on their apparent relationship to the stupa/granary form. In Dogon cosmology, Amma plays the role of the hidden creator/god. His first act, which we interpret in terms of the dual themes of the cosmology as the fertilization of an egg or the perception of a massless wave, is accomplished by himself alone, so it violates a principle of duality that defines the universe. It is thereby equated with masturbation, a sexual act that involves only one partner. According to the Dogon, one product of this impure act is the jackal, which in both Dogon and Egyptian cosmology represents the concept of disorder, so chaos is introduced into what was previously understood by the Dogon priests to be the perfectly well-ordered system of massless waves. Appropriately, one Egyptian word for jackal is *aasha,* a term that is also applied to the Egyptian god of destruction, Set.[12] Another Egyptian word for jackal is *auau,* a word that, like *aasha,* can also mean "to cry out."[13] In our view, the word *auau* represents the sound a dog makes when it opens its mouth in the shape of a <. By our interpretation, this is a concept that symbolizes the initial expansion of space when a primordial wave is perceived or the expansion of a womb after the fertilization of an egg. Likewise, the act of crying out can be seen as presaging the eventual utterance of a spoken word.

The Dogon priests tell us that the jackal, a likely counterpart to the Egyptian god Anubis, is symbolic of "the difficulties of" the surprisingly fallible creator/god Amma. From this point forward, the main impetus of creation would be to reconstitute a new order in the next world, the Second World, which is one stage beyond the Dogon

First World of matter that is thought to exist in a wavelike form. This catalyzed reordering of creation transpires within the conceptual boundaries of the Dogon Second World, which we take to be a likely counterpart to the Egyptian underworld. The ultimate efficacy of this new order is determined by yet another canine, this time the Egyptian god Sab, who is delegated to serve as judge between good and evil. He correlates to an animal that is both familiar to the Dogon and native to Egypt called the pale fox (*Vulpes pallida*), who in Dogon cosmology is similarly designated to act as judge between truth and error.

The pale fox (*Vulpes pallida*). *Based on WordNet 3.0, Farlex clipart collection. © 2003–2008 Princeton University, Farlex Inc.*

One troubling difficulty that we encounter when studying Egyptian cosmology is an apparent contradiction regarding the origin of the elemental deities. On one hand, through the filters of the theologies of various Egyptian cult centers, we are told how a creator/god initiates the emergence of the eight god/goddess pairs. On the other, it is also explicitly stated in important Egyptian texts that the great Egyptian mother goddess, Neith, gave birth to these gods and goddesses, who are therefore often referred to as Neters. Yet Neith plays no outward role in the events that surround their emergence within the context of the Ennead or the Ogdoad. How can this be?

An answer becomes evident when we consider these events from the perspective of the biological theme of the cosmology. If, on one level, these events symbolically describe the fertilization of an egg and the growth of an embryo, then we realize that, as both the Buddhists and Dogon suggest and the Egyptian word references affirm, these events must transpire inside a womb. It requires only moderate insight to infer that this can only be the womb of a mother goddess, such as Neith, who in Egypt personifies both major themes of the cosmology, since she is regarded both as the mother of the numberless deities known as Neters and as the weaver of matter. By this interpretation, Neith would actually appear as an essential player in the events described relating to the Ennead, even though she is not specifically credited. We, as observers, remain unaware of her presence because as the owner of the symbolic womb that is the egg-in-a-ball, she remains physically too large to fit in the picture!

Hereafter, from both a biological and a generative perspective, key events of the cosmology occur in pairs, much as the cells of a fertilized egg do when they begin to divide. From a cosmological perspective, these events begin with the formation of the axis munde, whose cardinal lines first bisect, then bisect again the egg-in-a-ball to produce four segments that symbolize the primordial elements of water, fire, wind, and earth and prefigure the formation of matter. Next, the eight master signs of the Dogon, counterparts to the Egyptian elemental gods and goddesses, emerge in pairs. The last of these, based on Dogon definitions, culminate in the formation of earth and sky, concepts that are symbolized in Egypt by the deities Geb and Nut. From a cosmological perspective, we interpret these events to represent the formation of mass and space.

The initial division of the fertilized egg, the act that corresponds to the "drawing up" of a perceived wave during the formation of matter as illustrated by the nummo fish diagram, creates a perfect twin pair, which constitutes the nummo of Dogon cosmology. In *Sacred Symbols of the Dogon,* we interpreted the word *nummo* as combining the Egyptian prefix *nu,* referring to waves of water, with the suffix *maa,* meaning "to examine or perceive." (From the perspective of a third creational theme relating to the formation of the universe from a primordial egg, the role of the perfect twin pair would likely be played by two atoms of hydrogen, the first element said to form in the big bang theory after the cooling of the quark-gluon plasma, and one that occurs naturally in pairs.)

From a similar perspective, we can understand the outer circle of the base of the stupa or granary—the figure that constitutes the ball of the Dogon egg-in-a-ball—in terms of the Egyptian phonetic root pet, which forms the basis of the words *pets-t*[14] and *pethan,*[15] meaning "ball." From the perspective of the transformation of matter, this ball constitutes the initial expansion of space that presumably occurs just after the initial perception of a massless wave. From a reproductive perspective, the ball represents a mother's womb, which expands along with a growing fetus. The first of these meanings is reflected in the Egyptian word *pet,* which means "to open out, to spread out, to be wide, spacious, extended." Our interpretation of this word depends on an understanding, first put forth in *Sacred Symbols of the Dogon* and supported in Buddhist stupa symbolism, that the square glyph □ conveys the concept of space.

TO OPEN OUT, TO SPREAD, TO BE WIDE, SPACIOUS (PET)

Definition of spaciousness	□	Space □ given, followed by the spaciousness glyph
		(see Budge, p. 255b)

This same phonetic value of *pet* can be seen as the root of the name of the Egyptian god Pteh—shown on the next page—whom Budge defines in his word entry (cited on page 91) as "the architect of heaven and earth . . . and the fashioner of the bodies of men." Appropriately, Pteh's role is also defined specifically in terms of the two major themes of the cosmology. Likewise, as we first explained in *Sacred Symbols of the Dogon,* a symbolic reading of the name Pteh can also be interpreted in the specific terms of those same two themes:

ARCHITECT OF HEAVEN, EARTH, AND THE BODIES OF MEN (*PTEH*)

The god Pteh	□ ⌓ ≬ 𓁰	Space □, mass ⌓, and DNA ≬, followed by the god glyph 𓁰 determinative (see Budge, p. 254b)

As we follow along with the architectural plan of the stupa or granary, what is initially conceived of as a circular base comes to be expressed conceptually as a three-dimensional hemisphere or dome ⌓. At this point, both the Dogon and the Buddhists compare the structure to a woman lying on her back, such that the hemisphere or dome, representing her rounded abdomen, takes on the position and appearance of an expanded womb. In the Egyptian hieroglyphic language, this notion is expressed by the word *kha-t,* meaning "womb," which can be written in its simplest form with a single glyph 𓄛, a glyph that Budge interprets to be a uterus. However, the word for womb can also be written in a more expanded form, one that, by our criteria, qualifies as a defining word for the hemisphere glyph ⌓. This longer spelling is written:

WOMB (KHA-T)

Definition of the the hemisphere glyph	𓄛 𓅐 ⌓	Uterus 𓄛 comes to be 𓅐, followed by hemisphere glyph ⌓ (see Budge, p. 570a)

Again, we can relate the word *womb* specifically to the concept of a shrine or temple by way of another entry in Budge's dictionary, which is written with the same uterus and hemisphere glyphs, carries the same pronunciation of "kha-t," and is defined as meaning "temple."[16] Likewise, we can correlate both of these concepts with the notion of nine emergent gods and goddesses who are likely counterparts of the Dogon god Amma and the eight Dogon ancestors through a third word entry pronounced "kha-t." Budge says that this word refers to "the body of the company of gods."[17] It is written with the uterus glyph and nine flags ; the flag is the traditional sign of a Neter. Budge defines the same word as meaning "of the body," defined as "issue, children," or the biological offspring that are the product of a womb.

An alternate name for the Ennead gods and goddesses, "the first and greatest nine gods," according to Budge, is *pestch-t.* Much like the word *kha-t,* Budge's first spelling of this word is written simply with nine flag glyphs. Based on a second spelling of the word, , Budge observes that "the true reading of the word is *Pauti,*"[18] which he describes as the name of a very ancient Egyptian god. The reading *pestch-t* is due to confusion of the signs , or *pestch,* and , or *paut.*[19] Budge defines Pauti as "the primeval god who created himself and all that is."[20] His name is based on the root *pau,* which we equated in previous volumes to the Dogon counterpart to the atom, the po. Budge defines the word *pau-t* as meaning "stuff, matter, substance, the matter or material of which anything is made,"[21] and describes a primeval god named Pau, whose name he surmises "perhaps means 'he who is,' or 'he who exists,' 'the self-existent.'"[22]

Likewise, the image of the paut glyph calls to mind a mythical episode at the beginning of Dogon cosmology in which the character Ogo, playing the role of light, steals a square (space) from Amma's placenta in order to form a universe. When we interpret the word *pestch-t* symbolically using Budge's proposed glyph substitution, thus transforming it to *pauti,* we find that it reads:

THE ENNEAD/NINE GREAT GODS (PAUTI)

Name of the nine great gods		Creation of the universe and matter , followed by three Neter flag glyphs (see Budge, p. 250b)

This word as Budge interprets it links the concept of the Egyptian Ennead gods and goddesses to the earliest phases of the formation of matter, expressed in distinctly Dogon terms, both through its pronunciation and through what Budge describes as the ancient concept of pau or pau-t, and by way of the specific graphic representation .

From the Dogon perspective, counterparts to the Ennead gods and goddesses are conceptualized as ancestors who are specifically correlated to segmented stages of the base plan of a Dogon granary. In regard to the corresponding component stages of a stupa, Snodgrass states that "an adequately designed building will embody meaning. It will express the manner in which the phenomenal world relates to the Real, and the One 'fragments into multiplicity. . . .'"[23]

In terms of the generative and biological themes of Buddhist stupa symbolism, this would refer to the initial division of cells that drives the growth of a fertilized egg into an embryo or the chain of events that presumably occurs to transform a perceived wave into particles. From this perspective, the initial figure, which includes the circular base of the stupa and its central gnomon (the raised portion of a sundial used to cast a shadow) , would represent the initial act of growth in each of those processes.

When we look to Egyptian words that express the concept of an ancestor, the first that we find, pronounced "tepi aui,"[24] begins with the glyph, which Budge defines to mean "foremost."[25] Our reading of the word *tepi aui* depends on the symbolic assignment of growth for the chick glyph, which we first proposed in *Sacred Symbols of the Dogon*. Based on that meaning, we symbolically interpret the word *tepi aui* to read:

ANCESTOR (TEPI AUI)

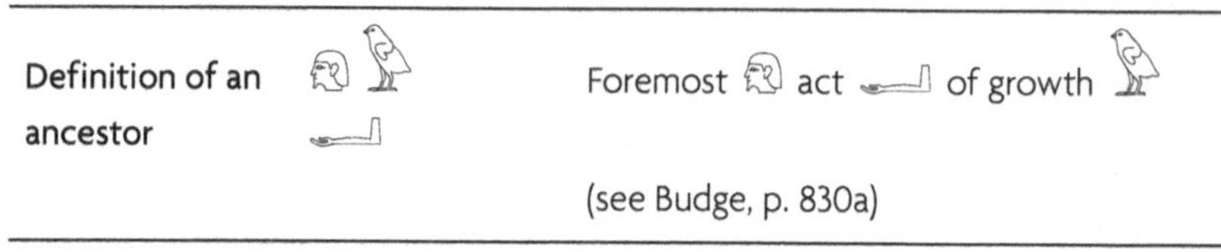

Definition of an ancestor		Foremost act of growth (see Budge, p. 830a)

The same three glyphs of this word provide the root characters for the word *tepiu-aui-ra*,[26] meaning "ancestors." This takes the form of a defining word for the sun glyph, which, of course, is the egg-in-a-ball shape. Thus, through the symbolism of Egyptian glyphs, the egg-in-a-ball shape is also correlated to the concept of ancestors. Symbolically, the word reads:

ANCESTORS (TEPIU-AUI-RA)

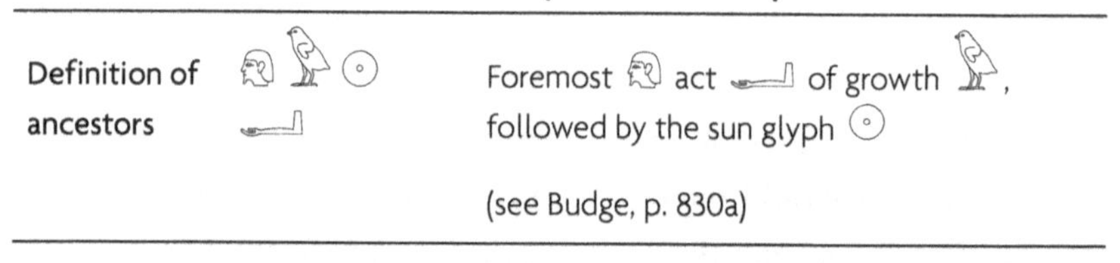

Definition of ancestors		Foremost act of growth, followed by the sun glyph (see Budge, p. 830a)

We can see based on the above examples that the ancient Egyptian concept of ancestors is formulated in relation to the same symbolic shape as the base plan of the stupa, that it symbolically expresses one key aspect of stupa symbolism, and that it defines in appropriate terms the concept of an ancestor using a figure that matches the Dogon cosmological references. If we take the Egyptian concept of an ancestor one step further, we can see that it also applies to the Ennead goddesses Isis and Nephthys, whom Budge refers to as "the two ancestresses."[27]

In terms of the biological theme of the cosmology, the concept of an ancestor is also expressed by the Egyptian word *pa,*[28] which forms the phonetic root of the words *pau, pau-t,* and *pau-ti,* and of the word *papa,* meaning "to bring forth, to bear, to give birth to." From both the generative and biological perspectives, the eight ancestors of the Dogon are portrayed as characters who are intimately involved in an initial expansion of space. In fact, the Dogon word *pa* means "to leave a space."[29] Not surprisingly, the Egyptian word *pa* reads symbolically:

ANCESTOR (PA)		
Definition of an ancestor	□	Space □ given , followed by the ancestor glyph (see Budge, p. 233b)

Dogon mythology introduces us to characters who do not appear in the Egyptian creation storyline as it has been passed down to us, but who nonetheless have likely counterparts in the Egyptian hieroglyphic language. One of these is the quick and impatient Ogo, who thinks he can create a universe as perfect as the one Amma has created. Ogo's actions, which are said to occur outside of the bounds of space and time and which Griaule and Dieterlen say actually prefigure the concepts of space and time, call to mind the scientific attributes of light. The likely Egyptian counterpart to Ogo is the Egyptian god of light, Aakhu,[30] whose name is written (light as a symbol) or (light deified). The likely relationship of Aakhu to the story of the emergence of the elemental deities is affirmed by the title *aakhuti,*[31] which is sometimes applied to Isis and Nephthys, two of the Egyptian Ennead goddesses.

In summary, we see that the plan of the aligned ritual shrine as it is understood in Buddhism and in Dogon cosmology defines a series of events that quite closely match the emergence of the elemental deities as described in ancient Egypt. Likewise, we find that we are easily able to affirm the key attributes of each deity—and often the specific relationship of one deity to another—as they were given in ancient Egypt in terms of the symbolic themes of the aligned ritual shrine. Furthermore, we are able to sensibly clarify Neith's likely role in creation in Egyptian mythology by interpreting it in terms of the stupa/granary plan that we propose for the parent cosmology.

NINE

THE CONCEPT OF THE PRIMORDIAL EGG

The concept of the primordial egg is one that appears almost uniformly in the cosmologies of widespread ancient cultures, and it can be taken as a signature sign of the parent cosmology we propose. If we know that a given culture assigns the creation of the universe to a cosmogonic egg, then we can presume it to be quite likely that we will find other evidence within the same culture of the influence of our parent cosmology.

Because of the multiple parallel symbolic themes of the cosmology as we have identified them, the egg is a symbol that can take on different aspects in different situations. It is therefore incumbent on the student of cosmology, when attempting to interpret symbolic meaning, to consider carefully the specific context in which a symbol appears. We have already made reference to the primary symbolic aspects of the egg during our discussions of the egg-in-a-ball and its likely relationship to the structure of matter and biological reproductive themes of the cosmology. However, there is another major perspective from which the concept of the primordial egg plays a key role, and that is in regard to the initial formation of the universe.

In Dogon cosmology, discussions of the formation of the universe are found primarily in surface-level myths, the public or exoteric folk tales that often bear suggestive qualitative resemblance to the creation stories found in the book of Genesis. It is perhaps *because* the concept of the primordial egg appears to have been so accessible to the general populace of noninitiates that it survives in such a ubiquitous way among so many ancient cosmologies. In the Dogon tradition, these myths serve to introduce and establish in a generic way many of the major themes and symbols of the cosmology. These stories describe the universe as having begun as a tiny ball of unrealized potential called Amma's egg. According to the Dogon priests, this ball, which housed all of the seeds or signs of the future universe, swirled and intensified while being seeded, yet somehow was also held in stasis for a very long time by two thorns. As the speed of the spinning egg intensified, it became more difficult to maintain equilibrium, so the egg eventually ruptured and released a whirlwind that scattered primordial matter to all corners of the universe.

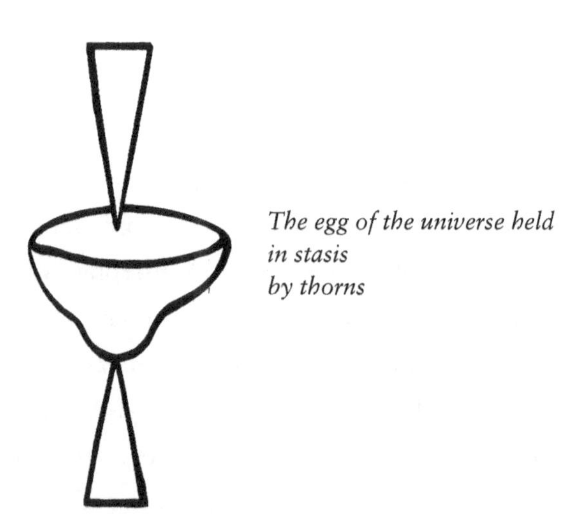

The egg of the universe held in stasis by thorns

Like everyday matter in its wavelike state, the Dogon priests consider the egg of the universe to have presented the potential matter of the universe in an ordered state. Dogon concepts relating to this cosmological view of the unformed universe are upheld in the Egyptian hieroglyphic language by words based on the phonetic root *ser* (written with the glyphs), which can mean both "to arrange, to order," and "little."[1] A phonetically related word, *s-rer,* is defined by Budge to mean "to make go round, to revolve," which is what the unformed universe did in Dogon cosmology.[2] One Egyptian word for thorn is *ser-t,* which is a defining word for the thorn glyph and is written:

THORN (SER-T)

Definition of a thorn		Arranged or ordered matter , followed by the thorn glyph (see Budge, p. 680a)

As we have mentioned, the bits of matter that eventually became the future stars and planets of the numberless galaxies of the universe are compared in Dogon mythology to pellets of clay, while the sun is compared to a clay pot that has been raised to a high heat, surrounded by a spiral of copper with eight turns. The moon is described as dead and dry, like dry clay. As the whirlwind of the universe emerged, the matter it released passed through stages of transformation that the Dogon priests compare to ice turning into water or water turning into water vapor. In science, these are called phase transitions. The immediate result of these transformations was the perfect twin pair known as the Nummo and the first finished creation of Amma, the po, which is a discrete unit of matter from which all other things are said to have been formed.

In a number of ways, the Dogon descriptions of Amma's egg serve to establish a rough equivalence between it and the Dogon granary, in both function and form. First, Amma's egg is defined as the repository of the seeds or grains of the universe, which is the same essential function that is served by the granary. Next, representations of Amma's egg in Dogon art depict its roughly conical shape, which can be seen as comparable to the shape of a Dogon granary. Finally, Amma's egg as it is understood in its role as the cosmic center from which all space evolved recapitulates the definition of a stupa or granary, which is seen as a symbolic representation of how multiplicity emerged from unity at the time of creation. Likewise, the very name of Amma's egg conjures a correlation between it and the Dogon picture of Amma that lies at the center of the egg-in-a-ball.

Among the key creational themes that are established by the Dogon rendition of the formation of the universe are the notion of an egg as the source of matter and the cosmos, the relationship of matter in its formative state to a whirlwind or spiral, the idea that matter comes to be transformed into particles (as represented by pellets of clay) and that these particles eventually combine in the form of clay pots (figurative receptacles of water), and the symbolic image that is overtly assigned to the sun during the course of the mythological storyline.

Given what we know about ancient cosmology, the choice of a common, quintessential symbol of reproduction such as an egg to symbolize the events of creation as they occur both in the macrocosm of the universe and the microcosmic world of vibrating threads can hardly be seen as fortuitous or coincidental. Rather, as we hope to demonstrate in later chapters of this volume, the symbol of the egg was more likely the product of very careful deliberation, aimed at orienting us from the very first presentation of Dogon myth to what is arguably the central theme of the cosmology, "As above, so below."

TEN

THE CONCEPT OF THE DIVINE WORD

For the Dogon, the very essence of creation rests with the notion of the formation of the Word.[1] This term as it is applied in ancient cosmology may well establish the earliest foundation for modern religious conception of "the Word of God." Like many other key terms of Dogon cosmology, our first impulse should be to view the many repetitive Dogon references to the term "the Word" as signaling the existence of another extended metaphor within the cosmology, one that is again meant to help guide us through the often tangled maze of symbols and meanings. This metaphor begins with the concept of a sign, as we have discussed it previously,[2] which is taken on one level as the earliest component stage in the formation of a spoken or written word. The concept of a sign is equated by the Dogon priests to a basic component of matter as we know it—the mythical counterpart to a fundamental particle. The notion of a spoken word is equated to matter after it has been reorganized by its journey through the Dogon Second World—the po pilu or the Calabi-Yau space—and has emerged after the piercing of the egg of the world. In the Dogon view, signs are intimately related to vibration, a concept that lies at the heart of both the notion of a spoken word as well as of vibrating threads and their presumed relationship to particles of matter. Conceptually, a string is thought to pass through a series of vibrations, each perceived as a different kind of particle. For the Dogon, as in string theory or torsion theory, these vibrations occur inside a primordial egg. As we have mentioned, the vibrations, which are characterized by the Dogon as seven rays of a star of increasing length, eventually grow long enough to pierce the egg. This act of piercing, which the Dogon consider to be both the eighth and culminating stage of a first egg and the initiating stage of a new egg, is defined as the conceptual point at which the finished Word is spoken. For both the Dogon and modern astrophysicists, these eggs in series form the membranes that constitute the woven fabric of matter.[3] Consequently, the process by which matter is formed is compared by the Dogon priests to an act of "weaving words."

In this context, it is easy to see that one purpose of the cosmological metaphor of the Divine Word may be to help orient us to think about matter as a product of vibrations. Likewise, one effective consequence of the metaphor is to associate in the mind of a Dogon initiate the very act of speaking with the processes of the formation of matter. Anyone who is familiar with the Judeo-Christian religious tradition is no doubt already familiar with this concept as it is illustrated in the book of Genesis, where the Hebrew God merely speaks a Word and thereby causes an act of creation to occur.

Supportive of the Dogon view of the Word and its symbolic relationship to the formation of matter is the Egyptian word *per,* which Budge documents as meaning "word" or "speech," a concept that he alternately defines as "what comes forth from the mouth." The symbolic meaning of this word depends on a root word that means "to go out" or "to go forth," which is written with the two glyphs [glyphs].[4] The first glyph is the character that, in *Sacred Symbols of the Dogon,* we associated with an Egyptian *arit,* the Dogon mythological counterpart to one of the wrapped-up dimensions of a Calabi-

Yau space, within which the vibrations of strings are thought to occur. Based on this association, we can see that the Egyptian concept of a spoken word appears to be expressed in symbolic terms that are distinctly similar to those of the Dogon.

WORD, SPEECH, WHAT COMES FORTH FROM THE MOUTH (PER)		
Definition of a spoken word		What comes forth from the mouth, followed by the spoken word/sign glyph (see Budge, p. 240b)

A second Egyptian defining word for the same glyph is pronounced "tchet" and is a homonym for the word *tchet,* which means "to speak" or "to say."[5] This root is also written with just the two glyphs, which we interpret to mean "given word," or more properly in English, "word given." This interpretation is based on discussions in *Sacred Symbols of the Dogon* that correlate the serpent glyph with the symbolic Word that emerges from the pierced egg during the formation of matter.[6] The word *tchet* takes the same basic form as the word *per* and consists of two glyphs that seem to define a third. It reads:

TO SPEAK, TO SAY, TO DECLARE, TO TELL (TCHET)		
Definition of the speech or word glyph		Word given, followed by the spoken word or sign glyph (see Budge, p. 913a)

We can confirm our interpreted idiomatic reading of the combined glyphs and of the speech or word glyph by looking to another entry in Budge's dictionary that means "to make a reply, to speak." This takes the unusual form of a defining word for both glyphs in combination. It is written:

TO MAKE A REPLY, TO SPEAK (ATCHET)

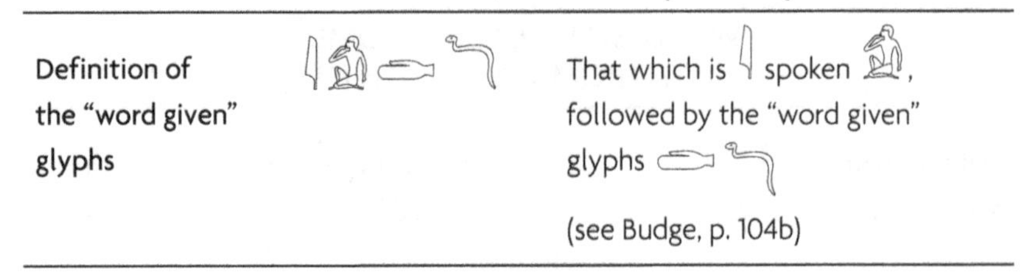

Definition of the "word given" glyphs		That which is spoken, followed by the "word given" glyphs (see Budge, p. 104b)

This same fundamental root, meaning "word given," forms the basis of the Egyptian word *tchet,* which according to Budge expresses the concept of the Divine Word or the notion of speech deified. Our reading of the word *tchet* depends on an interpretation of the falcon glyph that we discussed in chapter 5. In this discussion, we found that the Horus/falcon glyph can be taken to represent the concept of a symbol. This word *tchet* is written:

THE DIVINE WORD/SPEECH DEIFIED (TCHET)

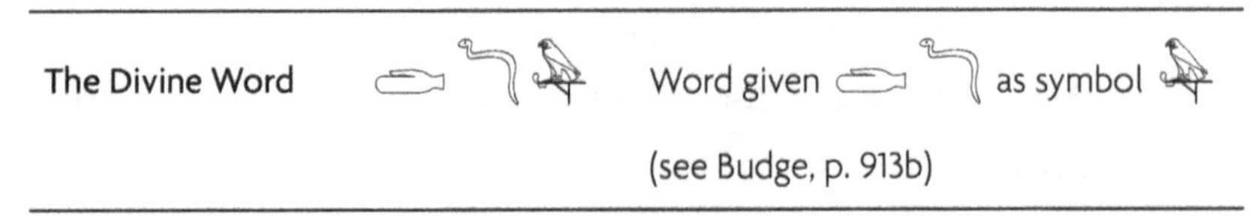

The Divine Word		Word given as symbol (see Budge, p. 913b)

We can see based on this example that the Egyptian concept of the Divine Word, like our view of ancient cosmology itself, is founded on the essential notion of words that are defined as symbols. From this perspective, we can interpret the Egyptian concept of the giving of the Divine Word as a ritual symbolic act, one that satisfies our first Egyptian definition of a symbol, as expressed in the word *ashem.* Divine Words, arguably the words of the cosmology, are those that are deemed to preserve knowledge in symbolic form.

An alternate view of the nature of speech is given by the Egyptian word *metu,* which Budge defines as "to speak, to talk, to say," and which can again be taken as a defining word for the same spoken word or sign glyph 𓀁. Symbolically, this dictionary entry seems to address the concept of speech in terms of a word's innate ability to enlighten:

TO SPEAK, TO TALK, TO SAY (METU)		
Definition of the speech or word glyph	𓌳𓂋𓀁	Perception or enlightenment 𓌳 given 𓂋, followed by the spoken word or sign glyph 𓀁 (see Budge, p. 335a)

It is interesting to note that, in accordance with the symbolic equivalence that is established in Dogon cosmology between the concept of words and the concept of cloth,[7] this word *metu* is derived from the phonetic root *met,* a word that means "cloth." The symbolic viewpoint expressed by the word *metu* gives us insight into the nature of the Egyptian hieroglyphic language as it may have been conceived by the ancient Egyptians themselves.

We find a direct reference to the Egyptian hieroglyphic language represented symbolically in one of the ancient names given to that language, expressed again in terms of the same phonetic root *met.* This name is *metut neter,* which Budge defines as meaning "words of the god" Thoth. The Neters, we recall, are the Egyptian gods and goddesses who were given birth by the great mother goddess, Neith, the weaver of matter. In fact, in *Sacred Symbols of the Dogon,* we specifically correlate these Neters to component stages in the formation of matter. So an alternate phonetic interpretation of the name metut neter, one that would bring it back conceptually to the Dogon sense of words as they relate to the formation of matter, might be "cloth of the gods who weave matter." The name metut neter is written:

THE EGYPTIAN HIEROGLYPHS (METUT NETER)

Description of the Egyptian hieroglyphs		Neter enlightenments , followed by the plural determinative (see Budge, p. 335b)

Our symbolic readings of these words reflect an Egyptian concept of the Divine Word that is defined first in terms of a word given as symbol. Here, Egyptian hieroglyphic words are defined both as words of the gods and as enlightenments, both of which imply, based on our understanding of the concept of the Divine Word, that they somehow carry symbolic import. Meanwhile, in Dogon cosmology we are encouraged again and again to think of words as being woven into the cloth. However in this case, the Egyptian term for words, *metut,* is expressed using a phonetic root that also actually means "cloth." Can it be that one purpose of this shared cosmological metaphor might really be to encourage us to look for precisely what we have found when we interpreted meanings for Egyptian words by substituting concepts for glyphs? What we uncovered is knowledge that has been effectively hidden in plain view through symbolic references that are virtually "woven into" the fabric of Egyptian words.

Having clarified the relationships between the predominant Dogon perspectives on the notion of a Word, the concept of speech, and the cosmological concept of the Divine Word, and the apparent relationship of each of these to ancient Egyptian words of similar meaning, the next task is to endeavor to tie these concepts back to Dogon cosmology based on other key cosmological words. I have spoken in earlier volumes about the Egyptian word *sba,* meaning "to teach,"[8] a related word for "pupil," *sba-t,* and their likely relationship to the Hebrew concept of a day of instruction called the Shabbat. Among the Egyptian words relating to the concepts of enlightenment and instruction, there is a group of words pronounced "am," and these words are related to a similar prefix in the name of the Dogon god Amma that we take to represent the concept of knowledge. The word *amma* in the Dogon language is both the name of the hidden god of Dogon cosmology—a likely counterpart to the hidden god Amen of ancient Egypt—and as we have pointed out, can also mean "to grasp, to hold firm, or to establish." (The Egyptian word *amen* can also mean "to make firm" or "to establish."[9] In correspondence with these Dogon definitions, one of the Egyptian words that is pronounced "am" also means "to know, to understand." This word is written:

TO KNOW, TO UNDERSTAND (AM)

To know, to understand		To come to know (see Budge, p. 6a)

The characters of this word again form the root of a second Egyptian word, also pronounced "am," meaning "to seize, to grasp," which in our view takes the form of a defining word for the grasping fist glyph . Even in modern-day English parlance, some five thousand years later, the notion of grasping something is still in common use as a metaphor for the act of coming to know it. This word *am* is written:

TO SEIZE, TO GRASP (AM)

To seize, to grasp		To come to know , followed by the grasping fist glyph (see Budge, p. 6a)

This same equivalence between the concepts of knowing or understanding and the phrase "to grasp'" is also expressed symbolically in another Egyptian word pronounced "am" that is written with an alternate set of glyphs, but still in the form of a defining word for the grasping fist glyph.

TO GRASP, FIST (AM)

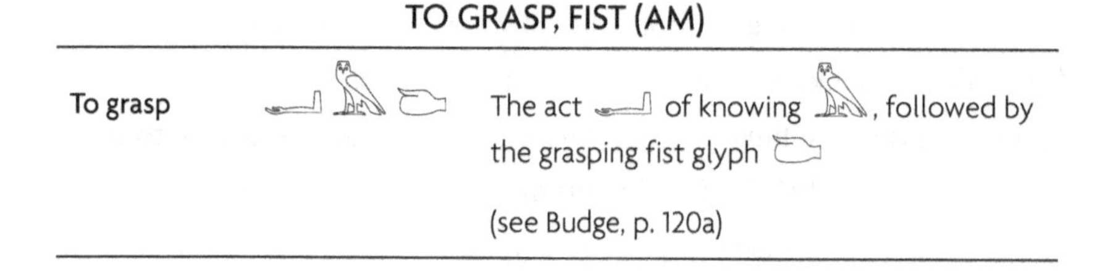

To grasp		The act of knowing, followed by the grasping fist glyph (see Budge, p. 120a)

Much as the above word appears to define the concept of grasping as "the act of knowing," another Egyptian word, *akhem,* meaning "to seize or to grasp," appears to define the act of grasping in terms of our familiar notion of acquiring knowledge. At the heart of this word is the root *akh,* which we associate with the concepts of light and enlightenment, and the word *am,* which we associate with the concepts of grasping and knowledge. This word again takes the uncommon form of a defining word for an idiom comprised of two glyphs in combination, , or the act of grasping.

DEFINITION OF GRASPING (AKHEM)

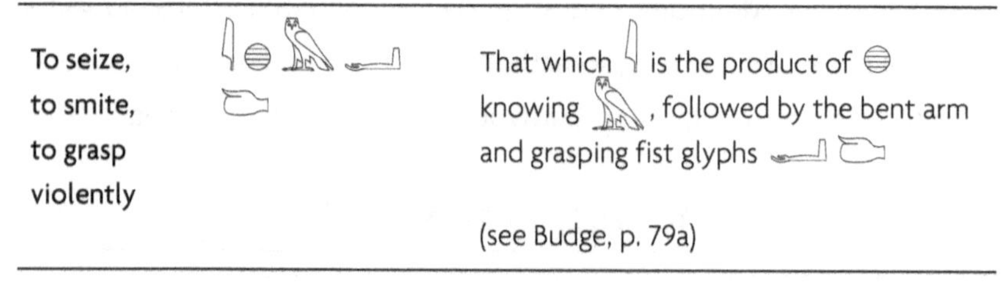

To seize, to smite, to grasp violently		That which is the product of knowing, followed by the bent arm and grasping fist glyphs (see Budge, p. 79a)

If, after these many related examples, we need yet another perspective on what the concept of grasping means, we can look to the Egyptian word *aam,* meaning "to clasp, to grasp, to seize," which appears to define what it means to come to know something. Again, this word takes the form of a defining word for an idiom that is expressed by two glyphs in combination , meaning "grasping's effect" or more properly in English, "the effect of grasping." This word is written:

COMING TO KNOW/THE EFFECT OF GRASPING (AAM)

To clasp, to grasp, to seize		The act of coming to know , followed by the bent arm holding a reed and grasping fist glyphs (see Budge, p. 111a)

ELEVEN

THE CONCEPT OF THE FISH

Throughout history, the fish, in a variety of forms, has taken on the role of a central symbol of many religions. In my opinion, the likely origins of this symbol are again best understood by comparing concepts relating to the fish in Dogon and Egyptian cosmology with those expressed in the Egyptian hieroglyphic language.

Taking the broadest view, the fish symbol is evoked in Dogon and Egyptian cosmology as part of a previously discussed ongoing organizational metaphor of the cosmology. Within this metaphor, as we have explained it, the component stages of matter are correlated to classes of creatures in the animal kingdom. These classes begin with insects, then progress to fish, then to four-legged animals, and finally to birds.[1] In keeping with the metaphor and the cosmological principle that creation is from waves or water, concepts of nonexistence and existence, which we associate with matter in its wavelike state, are symbolized by the dung beetle, which is classified by the Dogon as a type of water beetle. According to Calame-Griaule, the Dogon word for dung beetle is *ke,* which we take as a phonetic root of the Egyptian word for dung beetle, *kheper,* which is pronounced the same as the Egyptian words meaning "nonexistence" and "existence."

In accordance with this same organizing metaphor, we see remnants of insect symbolism that must have been originally assigned to the concept of the weaving of matter in the Dogon spider Dada, an arachnid, a type of creature that is also typically classified among insects. As we have mentioned previously, the Dogon word *dada* means "mother" and links us, along with other known spider symbols associated with goddesses in other cultures, to the Egyptian mother goddess, Neith.

The Dogon concept of the fish is expressed in terms of another key Dogon cosmological drawing called the nummo fish. The symbolism of this drawing is discussed at length in a chapter by the same name in *Sacred Symbols of the Dogon.* The figure represents a view of the egg-in-a-ball (the dark circle in the center of the figure) as it would appear if we took one giant's step back from it. As previously mentioned, this egg-in-a-ball carries both cosmological symbolism and symbolism of biological reproduction. The choice to illustrate concepts of biological conception in terms of a drawing of a fish is an interesting one, particularly in light of recent archaeological discoveries that suggest that the first in-body conception and reproduction in vertebrates occurred in fish.[2]

To quickly review, the figure can be interpreted as depicting the following images: (1) waves in the tail of the fish (which are defined as the underlying source of matter), (2) the point of perception of a wave that is the central egg-in-a-ball, (3) the dual figure of the pedestal on which matter in its wavelike form is said to be raised after an initial act of perception, (4) a squared hemisphere (the head of the fish), which we take to represent the concept of mass or substance, (5) and four whiskers, which represent the four quantum forces—gravity, the electromagnetic force, the weak nuclear force, and the strong nuclear force. It would be fair to say that, in Dogon cosmology, the fish defines the moment of perception of a wave, the same moment at which the future spiraling coil of matter is initially raised up.[3]

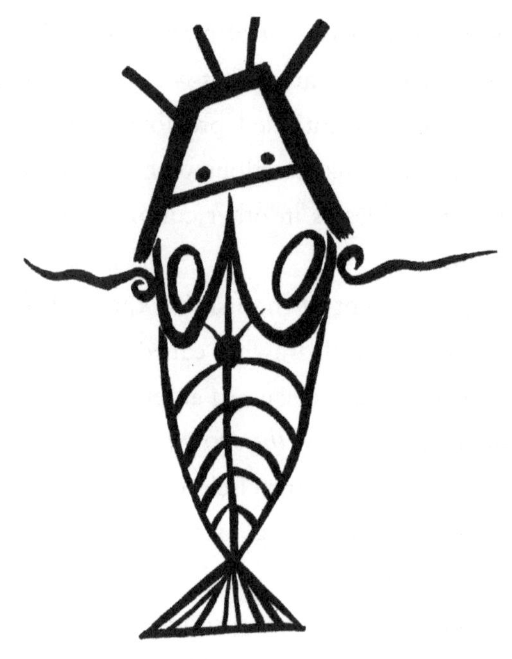

Diagram of the nummo fish

As our experience dictates, our first step when attempting to understand a Dogon cosmological term should be to look to Egyptian words that express the same concept. In this case, we find the likely symbolic definition we are looking for expressed in the very first of the many entries for the word *fish* in Budge's dictionary, which is pronounced "ata." We might interpret this word as a likely phonetic root of the name of the Egyptian creator/god Atum. The word is based on the Egyptian root *at,* meaning "moment"[4] and written 𓄿𓏏𓇳, which we take as a defining word for the Egyptian sun glyph 𓇳. Based on principles set forth in *Sacred Symbols of the Dogon* for reading Egyptian words symbolically, the leading glyphs of this word can be seen to define the final glyph. Among traditional Egyptologists, the sun glyph can represent a period of time, which is simply an alternately expressed definition of a moment. Based on this interpretation, we read the Egyptian word *ata* as another defining word, this time for the fish glyph .

DEFINITION OF THE FISH GLYPH (ATA)

A kind of fish		The moment at which that which is the spiraling coil of matter raises up and encircles , followed by the fish glyph (see Budge, p. 13a)

The symbolic concept of the fish is alternately assigned to another important cosmological shape in Buddhism and other later cosmologies. This is the shape of the looped string intersection , a figure that diagrams one of the three ways that strings are thought to interact with one another in string theory. This shape is also one that is evoked during the evolution of the base plan of the stupa and that is specifically identified in Buddhism (some three thousand years after predynastic Egypt) as the fish. It is recreated, somewhat like the famous optical illusion of the faces and the vase, in the space defined by the intersection of the two secondary circles that align the north-south axis of the stupa.

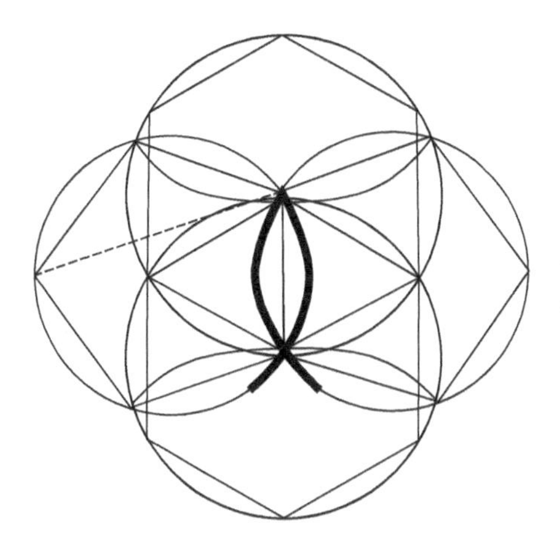

The fish image as seen in the base plan of the stupa

The second of these types of string intersection takes the shape of and is likely defined by the axis munde itself. The third type, , depicts a more complex interaction between two intersecting strings, can be said to be responsible for the metaphoric weaving of matter, and is found written as a glyph in the name of the Egyptian mother goddess, Neith, who is explicitly defined in Egyptian mythology as the weaver of matter. We interpret her name as a defining word for this glyph and interpret her name symbolically to read "weaves matter."

THE NAME OF NEITH (NET)

Name of Neith ( )		Weaves matter , followed by the complex string intersection and the goddess glyph determinative (see Budge, p. 399b)

While the Dogon nummo fish drawing has no known direct counterpart in ancient Egypt, the likely Egyptian counterpart to the Dogon word *nummo* supports our view of its meaning, which is the same as that explicitly described by the Dogon priests. We can interpret the Egyptian word to be the compound *nu maa*. This consists of two words, *nu*, which for both the Dogon and the Egyptians refers to waves and water, and *maa*, which means "to examine or perceive." These definitions support a view of the nummo fish drawing as representing what happens when a primordial wave of matter is initially perceived. Likewise, the central figure of the drawing, which the Dogon priests define as the egg-in-a-ball, defines the drawing's relationship to other concepts of creation and upholds this same interpretation.

As we demonstrated in *Sacred Symbols of the Dogon*, in terms of practical symbols as they are applied in the cosmology, the four-stage animal-class metaphor used to define the evolutionary stages of matter runs from insects like the dung beetle 𓆣 to fish 𓆛, to four-legged animals like the dog, fox, or jackal 𓃥, and to birds, where the flying bird 𓅮 is correlated to existence in its completed form, as given by the Egyptians word *pa,* meaning "existence," and the name Pau, the Egyptian god of existence. This would be the likely Egyptian counterpart to the Dogon po, or atom. The four-stage animal metaphor provides us with the information we need to interpret other references of the cosmology that are expressed in terms of these same four evolutionary stages. For example, the Dogon priests say that the earth is designated as the star of the fishes, a symbolic assignment that suggests that we should think of ourselves as having attained the second stage of development relative to concepts of civilization. Meanwhile, the Dogon priests associate the authors of the cosmology with Sirius, which is known throughout world mythology as the dog star. This suggests that we view these ancient instructors not as having attained the ultimate pinnacle of possible civilized evolution, but merely as having arrived at the third stage—somewhere above us, yet still substantially below the status of gods in our eyes.

In accordance with this four-stage progression of guiding cosmological references, the dung beetle, as an insect, is placed symbolically where it properly belongs, at the start of the progression. Next, we notice that the markings found on the back of the dung beetle, or

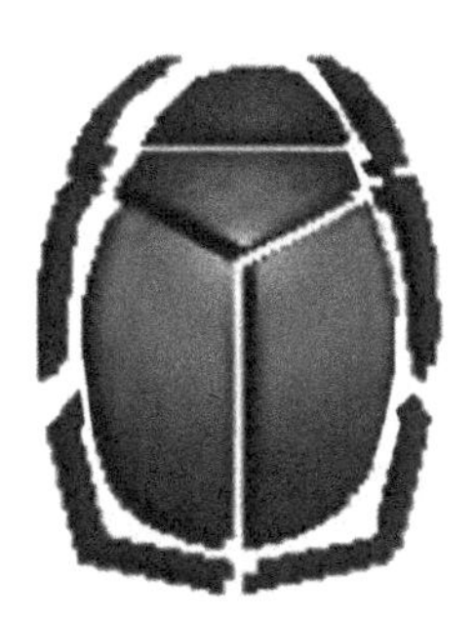

Markings on the back of a scarab; from www.ni-usa.com/construction/scarab/index.html.

scarab, roughly mimic the lines of the Dogon nummo fish drawing, which represents the starting point of the processes by which matter is created. These markings rise to the head of the scarab, which is represented by the shape of a hemisphere, the same shape that we interpret in Dogon and Egyptian symbology to represent mass or matter.

TWELVE

DEITIES

A few important facts relating to the concept of a deity come clearly into focus through the study of comparative cosmology. The first is that the parent cosmology cannot truly be said to have been a polytheistic system; in Dogon cosmology, only the creator god Amma rises to the status of an actual deity as the concept is understood in modern religious terms. Furthermore, as we come to a broader understanding of the rituals and symbols of ancient cosmology, we also realize that the concept of the worship of a deity has evolved greatly over time. In truth, our evidence suggests that the role of a deity in the earliest days of the cosmology may have fallen far closer to what we would interpret as a mnemonic device than that of an actual god or goddess in the traditional Egyptian, Greek, or Roman sense.

When we use Dogon cosmology, including the evolution of the stupa/granary form, and the defined characters of Dogon mythology as our primary guides, we find that we have been provided with several distinct examples by which to understand the concept of a deity. These begin with the Dogon creator/god Amma, whom, based on name, position within the cosmology, and principal acts credited to him within the cosmology, we take as a likely counterpart to the Egyptian god Amen and the ancient Hindu god Brahma. (Like

Amma, Brahma was traditionally seen as a creator/god with dual aspects—male and female—who was born from an egg that eventually formed the universe. Brahma is alternately described as having emerged from water.) The Dogon priests define Amma as the one true god of the Dogon religion, and they compare the concept of Amma to the force that facilitates the fertilization of an unfertilized egg or that initiates the formation of matter at the time of perception of a massless wave. This last definition is one that resonates with a symbolic reading of the Egyptian word *amen* as it applies to the concept of "the hidden one."

THE NAME OF AMEN (AMEN)

From the perspective of the biological reproductive theme of the cosmology, the stages of creation seem to be defined in terms of events that transpire inside a womb, so they imply the preexistence of a primordial Mother. For the Hindus, Brahma, like the Dogon Amma, was deemed to have been born without a mother, yet is still associated symbolically with the navel and the egg that define the womb of the Mother in the Dogon and Egyptian cosmologies. In Dogon cosmology, this Mother is the spider Dada, who is defined as the weaver of matter and whose name means "mother." In Egyptian cosmology, it is the great mother goddess, Neith, who is also assigned the role of weaver of matter. References to a comparable mother goddess are found in the early Vedic tradition among the Indus Valley civilizations, but it is thought that her attributes may have been absorbed into or reassigned to another deity in later times. Likewise, there are early symbolic associations with spiders in the Vedic tradition. For example, it is said that the Vedas compare creation to a spider's web that the spider first creates, then lies within. In ancient Egyptian mythology, the traditional concept of a god or goddess—a Neter—is specifically defined in terms of Neith, who is often described as the mother of all of the Egyptian gods and goddesses.

We have already discussed the first eight Neters in chapter 6, "The Egg-in-a-Ball." In Dogon cosmology, the likely counterparts to these Neters seem to be best understood as revered ancestors of the Dogon tradition who are associated with the concept of the eight master signs of Dogon cosmology. If the similarities between Dogon and Egyptian culture arise, as we believe, from extended contacts that may have occurred very early in Egyptian culture, then it is likely that the Dogon tradition preserves these concepts in earlier form.

In *Sacred Symbols of the Dogon,* we demonstrated a likely relationship between the Egyptian Neters and component stages in the formation of matter. Since the Neters are defined by their very name in specific relation to Neith, it would be hard to conceive of them as representing anything other than aspects or outgrowths of Neith. Such correlations would be a simple extension of the symbolic assignments previously outlined for the eight Dogon ancestors in relation to the egg-in-a-ball drawing. These ancestors relate symbolically to the various shapes evoked by the plan of the stupa, the very same shapes that we suggest may have been adopted as written glyphs in early Egyptian cosmological words. In other words, from the earliest stages of the cosmological plan, we see evidence of a likely three-way correspondence among (1) Dogon cosmological shapes and meanings, (2) Dogon ancestors and Egyptian deities, and (3) Egyptian glyph shapes and meanings. From this perspective, it is not surprising that we would find, as also demonstrated in *Sacred Symbols of the Dogon,* consistent relationships between the phonetic pronunciation of an Egyptian glyph shape and the name of an associated Egyptian deity.

Each glyph shape or cosmological drawing, which we believe at first symbolized a component stage in the cosmology, can thereby be correlated to an Egyptian deity whose traditional role in the cosmology defines in greater detail events that transpire at each component stage. One classic example of this kind of relationship is found with the Egyptian wave glyphs 𓈖 𓈗, which are pronounced "n" or "nu," and which we correlate to matter in its wavelike state, which is thought to be the primordial source of all matter. There is an Egyptian deified concept, named Nu, that Budge defines as "the primeval water whence everything came."[1]

Myths in Dogon society seem to have functioned primarily to introduce central concepts of the cosmology in a generalized way to members of the larger Dogon populace and, unlike the more familiar Greek myths, did not primarily report the acts of mythical heroes. However, myths take on a slightly different role in other cosmologies. In both the Egyptian and Buddhist cosmologies, these storylines often serve to define, through the actions and interactions they report, important identifying characteristics of Neter-style gods and goddesses. In my experience, when we get to the conceptual bottom of these myths, each can be seen as a kind of narrative metaphor that is used to define the concept or component stage of creation that is symbolized by the related deity.

Another class of Egyptian deity is typified by Anubis, the jackal-headed god of the Egyptian underworld. He is similar in type to the lesser-known Egyptian god of light, Aakhu, the likely counterpart to the mythological Dogon character Ogo. These deities, in my view, cannot be properly categorized among the Neters because it is clear based on Dogon references that they are more representative of cosmological concepts than of component stages of creation. Their counterparts begin simply as defined characters of Dogon mythology, and the actions of those counterparts within the myths serve to pinpoint precisely which cosmological concept they likely represent. Such characters include the Dogon jackal and fox, each associated with the Second World of Dogon cosmology. They include the unnamed twin sister of Ogo, from whom Ogo becomes separated and for whom he is eternally destined to search. This sister seems to play the role of time in the Dogon cosmological narrative. Similar themes are found in Hinduism, but expressed in the mythical relationship between Brahma and his daughter Sarasvasti.

Most students of cosmology understand the significant role that iconic objects can play when interpreting the likely symbolism of deities in some cultures. For example, traditional representations of Buddha, who we take as a likely counterpart of the Dogon po, or the atom, depict him holding or wearing objects that are known symbols of the cosmology. For example, it is not uncommon to see Buddha holding a representation of the Dogon egg-in-a-ball shape in his hand or wearing a string of pearls, a term that is applied by the Dogon priests to the po pilu, the eggs-in-series that weave the membranes, or branes, of modern cosmology. From a biological reproductive perspective, it is also common to find images of Buddha that depict him seated, with small children climbing on him. The Hindu god Shiva often holds a pot that is considered to be a kind of pitcher to hold water. Each of these objects can be taken as a clue to a more fundamental representative role played by deities. The same can be said to be true for deities in Egyptian cosmology—that the iconic objects they hold, wear, or are associated with have ultimate bearing on the traditional role that we interpret for each deity. These objects are often expressed in terms of shapes that are well defined in Dogon cosmology. Clearly, iconic objects bear a similar relationship to the deities of ancient Egypt. Each god or goddess is typically represented in conjunction with one or more of these objects.

A very similar view can be taken of the otherwise enigmatic tradition in Egyptian culture of depicting gods and goddesses with animal heads. The symbolic significance of specific types of animals and the roles they appear to play in the conceptual organization of Dogon cosmology all seem directly applicable to the Egyptian deities. For example, the scarab-headed god Kheper would correlate to the dung beetle, an insect symbolizing the first stage of creation in Dogon cosmology. In Egyptian cosmology, Kheper is considered by some sources to be the personification of the sun ☉, which is the image of the Dogon egg-in-a ball. In like fashion, Anubis, the god of the Egyptian underworld, is assigned the head of a jackal or a dog, an animal avatar that is appropriate to the Dogon Second World of creation. By the time we reach the conceptual top of creation in the Second World—the formation of the Word—we find the Egyptian god of language, Thoth, depicted with the head of a bird. In this way, the Egyptian convention of assigning animal heads to deity images can be seen as a way of marrying the concept of deities as component stages of matter with the various classes of animal that we believe were meant to serve as guiding references for organizing those component stages.

Another related aspect of Egyptian cosmology that is only suggested among the Dogon is the notion of the deified concept. These represent ideas that are so fundamental to Egyptian cosmology that they are treated with the same reverence as a deity and are so marked by the inclusion of specific determinative glyphs at the end of their written names. The closest likely counterparts to these in Dogon cosmology would be what we call cosmological key words, terms such as *nu, nummo, bummo, po,* and *sene* that can be seen as essential to the correct definition of Dogon cosmological concepts.

Once we recognize the three-way relationship that appears to have existed between cosmological concepts, glyph shapes (cosmological drawings), and Egyptian deities, then it becomes easy to understand why such a large number of glyph shapes (more than four thousand) and deities came to be the hallmark of Egyptian society. By our view, at least one new glyph and deity would have been required to express each new important concept. It would have been the ever-more-challenging job of the Egyptian priests and scribes to keep careful track of which shape represented which concept, and in association with which god or goddess.

Perhaps the single common thread that is to be found at the heart of each of these distinctive types of deities is that each represents, in symbolic form, some key aspect of creation. Again, experience tells us that if we are confused about the proper definition of a cosmological term, we should turn to the Egyptian hieroglyphic language for a succinct definition. The Egyptian word *neter,* which Budge describes as "the word in general use in texts of all periods for God and 'god,'"[2] can be written with a single glyph—the falcon glyph 𓅃. This is the very same character that we have previously suggested, based on Egyptian symbolic definitions, represents the concept of a symbol.

THIRTEEN

CIVILIZING SKILLS

As might be expected of a system of instructed civilization, Dogon cosmology as Griaule presents it is also focused on the transmission of specific civilizing skills. One or more of these instructed skills are correlated to each of the eight Words—comparable to lessons or sessions—of the larger civilizing plan. A strikingly similar mind-set is demonstrated in ancient Egypt cosmology based on the meanings of a family of words that are founded on the phonetic root *shes*. The Dogon concept of the Word itself seems to be expressed by the Egyptian word *shesher,* which means "utterance" or "speech."[1] This is founded on the same phonetic root as the Egyptian word *shesha-t,* which means "knowledge, skill, ability."[2] For the purposes of these discussions, these Egyptian terms will be taken as likely counterparts to their corresponding Dogon concepts. Symbolically, the eight civilizing Dogon Words can be seen to correspond to the eight stages of matter by which the po pilu or the Calabi-Yau space is defined.

In *Conversations with Ogotemmeli,* Griaule specifically defines many of the Words of the civilizing plan, each of which is considered to be fundamental to the very concept of civilized life.[3] Each skill (or related set of skills) is deemed important to differentiate mankind from the lower classes of animals that populate the earth. The introduction of the base plan of the stupa/granary structure, which was apparently used to establish basic concepts of mathematics, units of measure, and of units of time, can be seen as being a prerequisite to or existing concurrently with the introduction of the first of these civilizing skills. This suggestion is upheld by two Egyptian words pronounced "sesher," meaning "cord for measuring" and "to measure with a cord,"[4] and a phonetically similar word, *shep,* that defines the palm unit of measure.[5]

The first of the civilizing concepts described by Griaule, known as the First Word, is the concept of a simple fiber skirt. This concept "manifested on Earth the first act in the ordering of the universe."[6] It did so by imparting "words" of the "first language of the earth," which we interpret to have been the language of cosmology, or symbolism. Ogotemmeli, the blind hunter/priest who served as Griaule's designated instructor, compared the undulating fibers of this first skirt to the eight spirals of the sun, whose rays draw up the moisture of the earth. The likely corresponding Egyptian concept is reflected in the word *shes,* which means "cord, string, rope."[7] The purpose of speech that required this new language of Words (symbolism) was the organization, or more precisely, the catalyzed reorganization, of society, a concept that is also directly associated with the po pilu's role in relation to the formation of matter. The nummo fish, which represents the first stage in the weaving of matter (or speaking of the Word of matter) would be a conceptual counterpart to the fiber skirt.

Each Word or instructional stage of this organizing system carries symbolism that pertains to and in significant ways reflects the progression of creation within the cosmology. For example, the introduction of the fiber skirt, which we said constitutes the First Word, corresponds to the first stage of creation as it is defined in the cosmology. This stage relates to concepts of nonexistence and existence, so the Word was, appropriately, assigned symbolism by Ogotemmeli that is expressed in relation to waves and water and to the sun, which is the symbolic counterpart to the egg-in-a-ball drawing. These represent the conceptual starting point of the cosmology and can be seen as counterparts to massless waves in relation to the formation of matter. From there, key civilizing skills were introduced in a proscribed sequence, each one carefully correlated through its assigned symbolism to the next logical stage of creation. In this way, the instruction of civilized skills served to reinforce the overall plan of the cosmology.

For Griaule, the mere definition of each Word or instructed skill served as a kind of introduction to the symbolic concepts that are appropriate to the corresponding stage of creation. These were often given by Ogotemmeli specifically in terms of the biological and astrophysical themes of the cosmology. In accordance with the cosmological principles of duality and the pairing of opposites, each alternating skill presented by Ogotemmeli defined a class of activity that, within Dogon society, was alternately assigned, first to women, then to men. This became one of the ways in which the civilizing plan served to establish a working division of labor both for an individual household and for a broader human society. Similar divisions of labor appear to have been applied to society at large that might have served to foster the circulation of goods and services that would be required by a working economy.

Commensurate with the notion that each Word of the civilizing plan was meant to be reflective of a stage of creation, Ogotemmeli says that one consequence of the introduction of the First Word, like the initial act of perception of a massless wave, was that it resulted in a great deal of confusion and disorder among mankind. Ogotemmeli specifically compares this confusion to the disorder that is symbolically associated with the jackal when discussing the formation of matter.

The Second Word of Dogon cosmology relates to the concepts of spinning and weaving. Spinning, along with the notion fibers or threads, fell under the assigned domain of women, while the art of weaving fell within the provenance of Dogon men. Taken together, the two acts of spinning and weaving were equated symbolically by Ogotemmeli with an act of procreation (the first stage of creation), and he suggested that neither act by itself could produce anything that would be of value. As the second civilizing skill, the art of weaving was correlated to the second stage of the cosmology and to the divided egg-in-a-ball, which in its symbolic role as a womb reinforced the theme of procreation. Like the egg-in-a-ball, weaving was related symbolically both to the eight ancestors of the Dogon and to images of reproduction that call to mind the initial divisions of a fertilized egg, which is the second stage of creation in biological reproduction.

By this point in the instructional process, we can see that several broad metaphors of the cosmology have already been established. These include an implied, three-way equivalence among the concepts of weaving, the reproductive development of life, and the formation of matter. Also, we can see the establishment of a kind of identity between stages of creation and processes that relate to stages in the formation of a spoken word.

According to Ogotemmeli, the Third Word of the cosmology was the introduction of the eight grains of Dogon agriculture, along with a granary to house them. In Dogon culture, the harvesting and storing of the crop was often the responsibility of the women and children, while cultivation was the responsibility of the men. In the Egyptian hieroglyphic language, this instructional phase likely relates to the word *shes,* meaning "winnowed grain." It is at this stage of instruction that the symbolism of the stupa/granary form is first introduced, much of it in symbolic relation to the establishment of working classifications for the various types of animals and plants and in terms of the careful regulation of an agricultural cycle.

Symbolically, the Dogon granary served in a number of ways as a conceptual tool for world organization. Each face of the Dogon granary was associated with a star or star group whose risings and settings signaled the various planting and harvesting stages of the agricultural process. Consequently, skills of astronomic observation were a prerequisite to the establishment of a successful agriculture and to the establishment of a working agricultural calendar. The stupa or granary, whose careful orientation could now be understood as an essential civilizing skill, came to be a tool of astronomic observation in a second way, in tracking the movements of the sun and as an aid to implementing a calendar. Likewise, as we have suggested, each of the four staircases of the granary was associated with a class of animals. These were roughly characterized as insects, fish, domestic animals, and birds, the same four mnemonic symbols taken as Egyptian guiding references to four broad stages in the formation of matter in *Sacred Symbols of the Dogon.*

Another part of the Third Word of the cosmology, according to Griaule, was the introduction of drums and drum construction, which was considered to be a form of weaving. If the granary can be taken as a symbolic counterpart to the po pilu or the Calabi-Yau space in string theory, then this skill emphasized the importance of the vibration of threads in regard to this stage in the formation of matter. It also underscored both vibration and sound in regard to the formation of a spoken word. According to Ogotemmeli, what eventually came to be the most significant type of drum for the Dogon was covered with a chain of copper that rattled when played and changed the sound of the drum. This drum corresponds to the Egyptian *shesh,* meaning "sistrum," an ancient Egyptian percussion instrument that included a looped metal frame that rattled when shaken.[8] The introduction of the drum at this stage in the instructional process might be seen as an effort to blend a somewhat more recreational activity among the earlier, more taxing intellectual ones. The drum would be an important tool for attracting tribe members to rituals and festivals that would be part of the civilizing process, as well as an important practical tool for communicating messages across large distances.

The Fourth Word involved instruction related to the cultivation of the land, a process that was equated to weaving in that both were done by the same method. This instruction began with a system for dividing the land and, like the evocation of a stupa or granary, was equated with the notion of manifesting order from disorder, or purity from impurity. The process of plowing a field was equated to the formation of a Word, with the explanation that a plow moving across an uncultivated field was like a shuttle moving across the weft of a loom; it imparted conceptual words, or order, as it moved. This skill likely corresponds to the Egyptian word *shesa,* which means "to plow"[9] and to the Egyptian god *Shesa,* whom Budge defines as "a skilled plowing god."

The Fifth Word of the Dogon involves the introduction of clothing that has been made from the woven cloth. This same civilizing skill is likely reflected in the Egyptian hieroglyphic language by the word *shes,* which means "a garment made of linen."[10]

Although in *Conversations with Ogotemmeli,* Griaule is not always entirely explicit about his numbering of the conceptual Dogon Words as they relate to later civilizing skills, it appears that the Sixth Word concerned the introduction of metallurgy based on a forge that had been established on the roof of the granary as a conceptual part of the Third Word. This was to be used to produce the implements of agriculture that would be required for cultivation. In Dogon society, metallurgy was a skill that was assigned to men. It is in regard to the establishment of the forge that the myth of the ancestor who fires an arrow into the vault of the sky is related. It is also here that we find the tale of the Dogon ancestor who steals fire from the gods. The Egyptian words that correspond to this stage of the civilizing plan are *shesher,* meaning "arrow, spear, dart" and *shesher,* which means "to shoot."[11]

As Ogotemmeli tells it, while the forge was established on the roof of the granary, the grains themselves were to be housed in its eight inner compartments. Here we find a likely correspondence to the Dogon practice of establishing districts and villages in deliberate pairs called upper and lower. These were also meant to symbolize a proscribed division of labor among the Dogon whereby the makers of the implements of agriculture—the smiths—were forbidden to participate in the cultivation of the fields. Such a division of labor seems like a reasonable prerequisite to the establishment of a working economy.

The Seventh Word was the introduction of the art of pottery, a skill that falls within the domain of women. We have suggested in prior volumes of this series that the clay pot ⍥ symbolizes the concept of a particle, which we can think of as a kind of receptacle in which waves are collected and held. This symbolism is established through the public fireside myths that describe the planets and stars as pellets of clay thrown out at the time of the formation of the universe and the sun as a clay pot raised to a high heat. The likely related Egyptian words for this are *shes,*[12] which means "vase," and *shesh,*[13] meaning "vessel or pot."

If we remain true to the suggestion that Egyptian skills and concepts defined by the phonetic root *sesh* correspond to the civilizing skills introduced by the Dogon Words, then we have to consider the possibility that the Dogon Seventh Word also included instruction in how to bake bread, even though Griaule records no Dogon memory of this aspect of instruction. This possibility is raised by another of Budge's dictionary entries, pronounced "shes" and meaning "loaf of bread."[14] Not only would this be consistent with the next logical step of instruction in terms of the cultivating and harvesting of grains, but it also involves an application of heat similar to that used when firing a clay pot.

The Eighth Word constitutes the concept of language itself and correlates to the finished po pilu or Calabi-Yau space, which also represents the concept of mass in a completed state—mass fully woven or the Word of matter fully spoken and skills fully acquired. This aspect of the Dogon civilizing plan most likely relates to the Egyptian word *seshau,* meaning "intellectual abilities."[15]

Other concepts that suggest a close relationship between the instructed civilizing skills, or Words, to the cosmology include that of a blood sacrifice, which is a practice that is integral to Dogon society and closely related to the concept of the Word. According to Griaule, this practice relates to the flow of a circuit of energy that is thought to reside in and is transmitted through words. Ritual sacrifice—the spilling of blood—is considered by the Dogon priests as a way of giving energy back to the earth in a conceptual "completing of the circuit."[16] A corresponding Egyptian concept is suggested by the word *shesher,* which refers to an "animal for sacrifice."[17] Likewise, Dogon numerology and symbolic assignments to the eight Words pervade Dogon thought, so there are many different schemes by which the symbolic meanings of numbers play out, so many as to constitute a worthy study in its own right. Again, we can think of number symbolism as a conceptual metaphor by which intellectual groupings were intended to be organized. Divination is yet another skilled art that is practiced by the Dogon and that appears to bear an intimate relationship to the cosmology, and again could be made the subject of an entire volume. Last, but not least, is the art of written language, a skill that would appear to have been a logical and likely outgrowth both of any system of instructed civilization founded on the concept of Words and of this particular system of cosmology. The concept of written language will be addressed as the topic of the next chapter.

FOURTEEN

WRITTEN LANGUAGE

If we presume it to be true that ancient cosmology constituted an instructed civilizing plan, then those who may have implemented that plan would have faced some interesting difficulties, problems that would have called out for creative solutions. First of all, the nomadic tribes that are thought to have lived in Egypt at the time (somewhat prior to 3400 BC) would not have likely been familiar yet with the concept of a formal written language. In that event, some other reliable method of transmitting knowledge from generation to generation would have been required to perpetuate an ongoing instructed system. Interestingly, this is the same essential difficulty that my wife, Risa, and I faced early in our parenting days whenever we tried to take one of our preliterate toddlers to any public event. It occurred to us that, if through some unforeseen circumstance our children should ever become separated from us, they would have no way of communicating who they were, where they lived, or how to contact us. After some consideration, our solution to the problem was to teach our son and daughter their address and phone number in the form of lyrics set to the tunes of familiar children's songs such as the politically incorrect "One Little, Two Little, Three Little Indians." Once our children had memorized these new lyrics (not a difficult task for most toddlers), we simply instructed them to sing the words in response to certain leading questions such as where do you live? and what is your phone number? This approach worked well and turned out to be so effective that many of our adult friends from that time period, some having only heard the songs sung once, more than twenty years ago, are still able to recall our address and phone number based on the lyrics.

These kind of creative mnemonics, which were the answer to our family problem, seem also to have been the answer to how to transmit a complex cosmology from generation to generation without the benefit of written language. While our parental mnemonics played out as simple songs, ancient mnemonics as we understand them took a number of different forms. The most basic of these seems to have been the simple association of a concept or a skill with a geometric shape, such as the sun glyph shape ☉. Once successfully acquired, the habit of associating concepts with shapes would have been easily transferred to correlating a concept or life attribute with a specific living creature or object that may have fallen within the purview of the initiate's daily life. No doubt, ritual songs were also used to record and transmit important ideas or ritual processes, much as my wife and I used them with our children. Stories or myths were no doubt also a good way to introduce a more complicated series of thoughts or concepts to a bright initiate.

If these mnemonics were initially seen as a stop-gap method of transmitting information to be used only until a written language could be established as a civilizing skill among the initiates, they seem in the case of the Dogon to have ultimately proved themselves somewhat superior to written language in their ability to transmit whole—over a period of thousands of years and across many generations—the most intimate details of a complex cosmology. This assertion is upheld by the far greater level of detail that was recovered by Griaule and Dieterlen from the Dogon oral tradition, as compared with the patchwork of references that have often been recovered by researchers from the often badly fragmented written texts of many other ancient cultures.

When comparing the Vedic, Maori, or Chinese cosmologies with those of the Dogon and the Egyptians, it becomes clear that, in general, many of the specific cosmological words, especially deity names, could not have been strictly imposed as globally instructed components of the cosmologies. However, certain of what I call the base words and symbols of the cosmologies do turn up again and again in recognizable form. These include many of the most rudimentary words and symbols such as Amma, Amen, and Brahma, the concept of the po, and many of the most basic cosmological shapes and concepts evoked by the ritual aligned structure. But in any given culture, as we move further away from the concept of a creator/god and nondeified ancestors and closer to the concept of Egyptian- or Indian-style gods and goddesses, most cosmological terms seem to consistently reflect the language of the individual culture in which they are found. This suggests that these words of the extended cosmology could not have been predefined terms overlaid on the culture through instruction, but rather were later ones that emerged sometime after the initial presentation of the civilizing plan.

According to our proposed chronology for the implementation of the civilizing plan in any culture such as Egypt, the first symbolic characters of written language seem to have been adopted from shapes and concepts that are evoked by the stupa/granary plan, along with various drawn shapes used to support the definitions of various cosmological terms. These shapes/concepts were sometimes applied in the Egyptian hieroglyphic language individually as single-glyph words, such as the word for "day" or "sun," written with the sun glyph ☉, or combined as multiple-glyph shapes, such as the Egyptian word for year ʃ ☉ (which we interpret symbolically to read "the time of the earth's orbit around the sun").[1] These produce discrete symbolic definitions that, in my view, constitute the essence of many Egyptian cosmological words. This outlook is a sensible one, given that the Dogon possess both the spoken cosmological words and the drawn shapes/concepts, but do *not* apply these shapes as actual written words.

While conducting my initial research into Dogon cosmology, as I familiarized myself with the symbolic shapes and meanings of ancient cosmology, it was hard not to notice ongoing similarities to the shapes and meanings of ancient Egyptian glyphs. It was also hard to ignore the somewhat confusing fact that the Dogon priests seemed to have retained a fluent grasp of each cosmological shape and meaning, yet had never acted on an impulse to apply these as actual written characters. Given the many intimate cultural, religious, and linguistic parallels that we have observed to exist between the Dogon and the ancient Egyptians, what could this surprising absence of a written language among the Dogon mean?

I felt that surely it was not believable that a culture such as the Dogon, who are known to place great importance on preserving intact the significant details of their own tradition, to say nothing of their reverence for the purity of words,[2] could have once possessed and then somehow, altogether as a society, completely suppressed any memory of their own system of writing. Nor is it believable that they could have done so yet still retained many of the key images and traditionally assigned meanings that so often closely resemble Egyptian hieroglyphs. Then I wondered whether it could be possible that the implied relationship that seems to have existed between the Dogon and ancient Egypt could have occurred sometime prior to the first appearance of written language in Egypt. In that case, it would be possible that the Dogon might have ended their contact with ancient Egypt sometime before the concept of writing had actually came into use, so the Dogon priests might never have come to be familiarized with writing as a concept.

Discussion in *Sacred Symbols of the Dogon* shows that there is supporting evidence to suggest a very early connection between the Dogon and predynastic Egyptian cultures.

It should be noted that there is a large body of glyph shapes, given in the form of animals, plants, and dancing figures, whose likely evolution can be traced from predyanastic Naqada pottery to later written Egyptian glyphs, and which are often cited as proof of the slow evolution of written language. These figures have little bearing on our discussions because they rarely, if ever, appear in the cosmological words whose pronunciations and meanings we correlate between the Dogon and ancient Egypt. For the purposes of this study, our primary concern is with cosmological symbols and words, how they may have entered written language, and how they may be used to convey meaning in written Egyptian words.

It is here, however, that we encounter an apparent contradiction. Written Egyptian glyphs, whose historic pronunciations are uncertain, are often assigned one or more phonetic values by traditional Egyptologists, whereas the Dogon cosmological shapes have no direct phonetic associations, at least not individually. For example, the Dogon drawing of the sene seeds is not considered by the Dogon priests to be a phonetic symbol that carries the pronunciation "sene." In our view, the Egyptian glyph shapes used to write a given word must have been selected—first and foremost—based on the concepts they represent. The phonetic value of the Egyptian word would then appear to be a secondary effect, derived as a consequence of the glyph shapes that were ultimately chosen to write it. So how can it be that the Dogon priests, who have no apparent knowledge of the written words, pronounce their cosmological words according to the Egyptian schema?

The likely solution to this conundrum lies in what I call the mind-set of the cosmology. The very capable efforts that went into the apparent design of the cosmology are uniformly evident, first in the plan of the mnemonic stupa/granary form, and next in the complex parallel themes that are so remarkably represented using a single, shared set of symbols. As we demonstrated at length in *Sacred Symbols of the Dogon,* this expansive set of symbols as defined within the Dogon oral tradition is a close match for a corresponding set of written Egyptian glyphs. The suggestion is that the concept of writing as a civilizing skill was a part of the instructional plan as it was originally conceived, before its instruction actually played out in Egypt. Within this plan, each key cosmological concept may well have already acquired its eventual phonetic pronunciation long before the skill of writing was introduced. This conclusion makes sense, since any given cosmological concept, such as the po or the sene, would have required a proper spoken name to effectively play its role in the Dogon oral tradition. This same solution would also explain the very specific conceptual assignments that have been demonstrated to correlate to Egyptian phonetics as stand-alone values.[3] From this perspective, the simple pronunciation of a given sound would be the conceptual equivalent to the act of writing a character to represent that sound. For example, it appears that the sound "s" implied the concept of bending or binding, "t" implied the concept of mass or matter, "p" correlated to the concept of space, and "a" represented the concept of existence. By this scheme, the word *stupa,* which combines the phonetic values "s," "t," "p," and "a," would represent a symbolic sentence meaning "the bending or binding of mass or matter into existence."

In English, letters represent phonetic sounds and words represent concepts. We portray a word in English by either speaking the sounds or writing the letters that replicate the sound of the spoken word. To form a sentence, either verbally or in writing, we combine appropriate words together based on the concepts they represent. We read or interpret this sentence by mentally substituting preagreed concepts for these words.

In our view of the Egyptian hieroglyphic language, at least as it applies to words of ancient cosmology, each drawn glyph represents a concept. To form a written word, we select the glyph shapes whose associated concepts best express the word's intended meaning, then list those shapes together to form a kind of symbolic sentence. This sentence, by definition, expresses the meaning of the word. In effect, this process takes the notion of meaning one level deeper than we are accustomed to in relation to the English language model. In English, the letters "m", "o," "n," "t," and "h" tell us nothing whatsoever about the concept of a month, whereas the glyphs for the Egyptian word *abt,* meaning "month," (symbolically, "the moon makes an orbit") appear to specifically define the very concept that they represent.

One advantage to this approach to reading Egyptian hieroglyphic words is that it offers a coherent explanation for the frequent inclusion of unpronounced trailing glyphs in various ancient Egyptian words. By our view, the leading glyphs of the word serve to define the symbolic concept that is to be assigned to the trailing glyph. We refer to this kind of symbolic word as a defining word.

From the perspective of the Dogon, who have no written language and who define matter as being derived from primordial vibrations that form conceptual words, it is clear that the concept of the Divine Word refers to the notion of a spoken word, not to one that has been written. Likewise, the various Egyptian words we have explored that relate to the concept of a spoken word appear to be founded on similar phonetic roots and root concepts as the Dogon Word. Given this, it seems reasonable to conclude that the Divine Word of the Dogon and Egyptian cosmological traditions, like that of the often-similar

Judeo-Christian religious traditions, is conceived of as a word that is spoken aloud by a creator/god. (The Hebrew God did not write, "Let there be light!") This would suggest that the first written words might have constituted a kind of symbolic extension of a spoken word and that the concept of written language might well not have appeared until sometime after the first appearance of ancient cosmology as an oral tradition, much as we now see it to exist among the Dogon.

Once again, if we want to understand the concept of the written word and what it may have meant to the ancient Egyptians, it makes sense to focus on various Egyptian hieroglyphic words that relate to the concept. As we begin to explore these words, we first notice that they are formulated in ways that are symbolically similar to the spoken-word examples given previously. Just as the Egyptian spoken-word definitions appear to have been conceptualized in terms of glyph idioms such as *tchet,* meaning "word given," and *am,* meaning "to come to know," likewise Egyptian words that express concepts of writing that seem to center on the notion of "placing" or "setting" an already-spoken thought into writing. We mentioned in *Sacred Symbols of the Dogon* that Budge defines an idiom *uti,* meaning "to put, to place, to set," that is written with the two glyphs .[4] These form the symbolic root of the word *ut,* which means "to write, to inscribe, to engrave."

TO WRITE, TO INSCRIBE, TO ENGRAVE (UT)

To write, to inscribe, to engrave	To place in writing (see Budge, p. 190b)

Budge also defines a phonetically related word, *uten,* meaning "to copy, to write," that is based on the same glyph idiom and that takes the form of a defining word for the writing hand glyph. Symbolically, the word *uten* reads:

TO COPY, TO WRITE (UTEN)

To copy, to write		To place in writing, followed by the writing hand glyph (see Budge, p. 191b)

The concept of the written word as a formalized extension of the spoken word may be illustrated by the name of the Egyptian god Uteb, whose precise role in Egyptian mythology Budge does not define. Some translations of *The Egyptian Book of the Dead* identify Uteb with Osiris, the same Egyptian god that we associate with the civilizing plan of ancient cosmology itself.[5] However, we can see that the name Uteb is written using the same glyphs that we assign to spoken and written language, so a likely relationship to those concepts is reasonably inferred. Our symbolic reading of the name Uteb depends on an understanding of the glyph, which was discussed previously, as representing the concept of a spoken word. The name reads:

NAME OF AN EGYPTIAN GOD (UTEB)

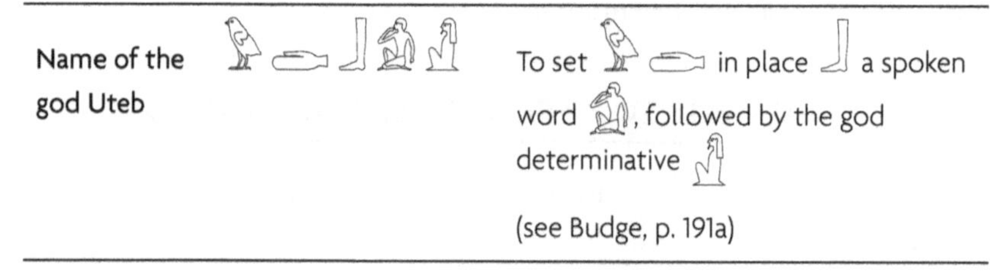

Name of the god Uteb		To set in place a spoken word, followed by the god determinative (see Budge, p. 191a)

Based on this symbolic reading, we can tentatively identify Uteb as an Egyptian god whose specific role may have been to oversee the placement of words into writing. This would imply a kind of deliberate and self-conscious approach to the implementation of written language that is consistent with what we know about ancient Egypt, where formalized writing did not appear until sometime after the earliest days of the civilization. Likewise, it is consistent with our symbolic method of reading words, which implies careful consideration on the part of the writer as to the choice of appropriate glyphs to convey discrete concepts.

Within this mind-set of language, ancient cosmology as it appears to have been sustained among the Dogon and in Buddhism can be seen as a likely precursor to—and perhaps even an active source of—written language in Egypt from a number of different perspectives. The first of these involves the simple "placement into writing" of cosmological concepts based on definitions and drawn shapes that are similar to Dogon cosmological definitions and drawings. In fact, our symbolic method of reading Egyptian words constitutes a kind of mirrored counterpart to the act of placing a spoken word into writing. Theoretically, to write an Egyptian word symbolically, a priest or scribe would have carefully considered the concept or thought that he wished to express, then selected appropriate symbolic shapes that would succinctly reflect that thought. Conversely, for us to read that same word, we must consider the various symbolic meanings that can accrue to each glyph shape, then settle on the likely ones that apply to each word as we try to carefully interpret the symbolic sentence that will result.

A remarkably similar process of encoding and decoding symbolism plays out again and again in the daily life of the Dogon, for whom, according to Griaule and Dieterlen, most acts of daily life and many of the objects within their daily environment carry specific symbolism that may call for explanation. Likewise, the many specific drawn symbols of Dogon cosmology are so intimately associated with their related concepts that Ogotemmeli once told Griaule that he typically felt compelled to draw an important concept in order to explain it properly. Thus we find among the Dogon nearly all of the apparent prerequisites for writing an Egyptian word symbolically, including similar drawn shapes associated with similar concepts, which are routinely encoded and decoded for purposes of discussion, as one might do when writing or reading an Egyptian word.

We can substantially affirm the notion that Egyptian words were meant to be read symbolically by looking to Egyptian words that expresses the concept of reading. Our symbolic translation of the first of these, pronounced "sha," depends on our interpretation of the divided river glyph, which was set forth in the discussion of symbolism in chapter 5, as presenting the image of a dammed river, such as would be required to create a reservoir. In the word *ashem,* meaning "symbol," we interpreted this same glyph to reflect the concept of preserving symbolic meaning. The notion that divine words preserve meaning is one that is specifically upheld by Budge in his definition of the word *ukhet,* meaning "treated with drugs, embalmed." He states in his dictionary entry that the same terminology is "also used of words of the wise which are 'preserved' or stored up."[6] From this perspective, the word *sha* reads:

TO READ (SHA)

To read	That which has been preserved comes to be spoken (see Budge, p. 722b)

It may be significant in terms of the relationship of drawn symbols to Dogon and Egyptian cosmology that the pronunciation "sha" also forms the phonetic root of the Egyptian word *sha-t,* which Budge defines to mean "something decreed or ordained by God, what is ordained by man or fixed by custom."[7] The same phonetic root forms the basis of the name of Shaait, the goddess of primeval matter, the word *shaa,* which Budge defines as "the source of life," and the word *shaa,* which refers to a "warehouse, storehouse, or granary."[8] In these ways, the concept of reading appears to be intimately entwined with many of the key themes that form the basis of Dogon cosmology.

The resemblances that we often observe between Dogon cosmological drawings and Egyptian glyph shapes suggest that early written glyphs may have taken their form and meanings from cosmology. This suggestion is clearly illustrated in terms of the Egyptian sun glyph ☉, which traditionally carries the meaning of "sun" or "day" and which appears in words whose meanings relate to periods of time. This shape, in association with these same symbolic meanings, is already familiar to us based on the plan of the aligned ritual shrine. As we have mentioned, the circular base, which both the Dogon and Buddhists agree symbolizes the sun, acts first in the role of a sundial to help an initiate track the hours of a day, then as a tool to measure the length of a year, determine the dates of the equinoxes and solstices, and follow the movement of the seasons. So in terms of Dogon and Buddhist cosmology, both the shape and its meanings as traditionally given for the Egyptian hieroglyph are completely understandable.

The same can be said for other Egyptian glyphs whose shapes bear direct resemblance to component shapes of the stupa/granary structure. For example, the square, flat roof of the Dogon granary that is said to represent empyreal sky, which we equate to the concept of space, finds its shape and meaning reflected in the Egyptian square glyph □, which appears in words whose meaning reflect concepts of spaciousness. Likewise, we suggested in chapter 7 that the shape of a hemisphere, which in Dogon and Buddhist cosmology is taken to represent a womb, appears to be defined by an Egyptian word that means "womb." So we have demonstrated an almost predictive relationship between symbolic figures that are evoked by the aligned ritual shrine and comparable Egyptian glyph shapes and meanings.

This pattern of relationship between Dogon and Buddhist cosmological shapes and Egyptian glyph shapes and meanings is not limited to forms evoked by the stupa/granary structure. In fact, it seems to extend to many of the cosmological shapes that are drawn by the Dogon priests to explain varied facets of their cosmology. For example, in previous volumes we have mentioned the shape of the Dogon nest drawing, which is used to illustrate a component of matter comparable to an electron, and the relationship it bears to the nest glyph as it appears in various Egyptian hieroglyphic words. In general, where the Dogon priests define a symbolic object such as a clay pot or a spiraling coil, or a cosmological concept such as water, fire, wind, or earth, we find Egyptian glyphs to express the same concepts in similar terms within various Egyptian hieroglyphic words.

Another aspect of the Egyptian hieroglyphs that strongly suggests that their original purpose could not have been primarily phonetic is the sheer number of glyphs. Most modern languages seem quite able to represent a wide range of phonetic sounds using only a relative handful of written characters. If Egyptian writing had been driven by an impulse to record words phonetically, it seems the priests would have reached a point where there would have been no need to continue inventing new phonetic glyphs. In my opinion, the simple existence of literally thousands of drawn characters contradicts the idea that they could have been originally meant to serve as phonetic placeholders. From an opposing perspective, any system based on the pairing of concepts with drawn characters might well require a much larger number of glyphs.

The clear suggestion is that the earliest written Egyptian glyphs may have taken both their shapes and meanings from preexisting shapes/concepts defined in Egyptian cosmology. Evidence of a likely relationship between the Egyptian concept of a shape and the Egyptian creation tradition is found in the word *kheperu,* which Budge defines to mean "form, manifestation, shape, similitude, image."[9] In this word, the very concept of a shape is defined in terms of the kheper, the Egyptian dung beetle that symbolizes the concepts of nonexistence and existence.[10] If at this point we are still in doubt about the association of concepts of existence with Egyptian glyph shapes, we need only look to the related word *kheperu,* meaning "a pot," which assigns the very same phonetic root to one of the quintessential symbols of both the cosmology and Egyptian writing.

One point to keep in mind when pondering the meanings of Egyptian glyph shapes is that an ideal symbol can be seen to draw its meaning from nature. We can illustrate how this point plays out in terms of the shapes of the stupa/granary form by considering the shape of the sun glyph, which can be defined in several ways by movements of the earth in relation to the sun. The circular sundial shape is the product of the progressive shadows of the central gnomon that are cast by the sun as the earth rotates. The east/west line, whose back-and-forth motions delineate the solstices and equinoxes, defines this same circular shape in relation to the earth's movements during its orbit around the sun. From this perspective, the likely symbolic relationship of the sun glyph to a meaning in the Egyptian hieroglyphic language is clearly illustrated by the table of words and glyphs on the next page, whose definitions relate to units of time or to movements of the earth in relation to the sun.

THE EGYPTIAN SUN GLYPH

If we read Egyptian words ideographically—interpreting each character as a symbol for a defined concept—we can show that the sun glyph ☉ actually depicts an orbit and on one level symbolizes the orbit of the earth around the sun. Budge defines the outer circle ○, as meaning "to encircle, to orbit" (See the word *shenu,* Budge, 743b).

By that interpretation, the dot at the center of the circle would be the sun.

Year		The time of the earth's orbit around the sun (See the word *renp-t*, Budge, p. 427b)
Month		The moon makes an orbit (See the word *abt*, Budge, p. 40b)
Seasons		The three bends/arcs of the earth's orbit around the sun, followed by the number three determinative (The ancient Egyptians observed a three-season year.) (See the word *ses*, Budge, p. 696ab)

On another level, the sun glyph also symbolizes the rotation of the earth in relation to the sun, or a day.

Day		Rotation of the earth in relation to the sun (This word is also written with the single glyph, so it comes to symbolize the concept of a day.) (See the word *hru*, Budge, p. 450a)
Week		Rotation of the earth in relation to the sun ten times (The ancient Egyptians observed a ten-day week.) (See the word *met*, Budge, p. 331a)
Dawn		Caused/created by the rotation in space of the earth in relation to the sun. (See the word *nehp*, Budge p. 381a)

Since each of these fundamental units or stages of time is defined in terms of the earth's relation to the sun, the sun glyph comes to symbolize a unit of time.

Based on Egyptian hieroglyphic word examples such as these, in *Sacred Symbols of the Dogon* we outlined a new approach to reading Egyptian hieroglyphic words. We proposed that this method applied first and foremost to words that relate to the cosmology. From this perspective, we interpret each Egyptian word as a symbolic sentence composed of drawn shapes, each of which represents a well-defined concept.

FIFTEEN

SYNCHRONIZING COSMOLOGIES

The Na-khi-Dongba of China

By this point in our studies, we should have acquired a firm grasp of the societal elements that constitute signature signs of our proposed parent cosmology as we understand it, based on the Dogon, Egyptian, and Buddhist models. If, as the Dogon and Buddhists claim, ancient cosmology represented an instructed civilizing plan—one that was deliberately implemented, and perhaps customized or tailored to various extents, in many different regions of the world—then the appearance of one or more of these signature elements within the context of some other very ancient culture should serve to spark our intellectual curiosity.

We know based on our research into the nature of aligned ritual shrines that the stupa is a traditional form of shrine that is commonly found throughout India and Asia, so its appearance within a culture—say, such as China—would strongly suggest the early influence of this same parent cosmology. Unfortunately, few structures except those built from stone are likely to survive intact for more than a few hundred years, so we might not expect to ever find a four-thousand-year-old wooden stupa, even if one had, in fact, existed. Likewise, the earliest written records in China date back only as far as the early Shang dynasty (which began around 1700 BC), in the form of characters that were carved into Chinese oracle bones. Consequently, we have little firm written evidence prior to that time or from that region of the world from which to infer the presence of any of the signature elements of the cosmology.

We do, however, have strong circumstantial references that point to the familiar influence of this same cosmology. The first of these comes out of a text by Paul Wheatley of the University of Chicago called *The Pivot of the Four Quarters,* which is a preliminary study of the rise of the earliest civic centers in China. According to Wheatley—and also according to Snodgrass, who quotes Wheatley in his book *Architecture, Time and Eternity*—ancient cities in China were uniformly aligned to the cardinal points using the very same method of alignment as is described for a stupa by Snodgrass in *The Symbolism of the Stupa.*[1] Furthermore, Wheatley, speaking in respect to ancient Chinese cities, agrees with an observation about Indus Valley settlements made by anthropologist Walter Fairservis of Vassar College, who suggests that the earliest ancient cities grew up not around centers of agriculture or finance, but rather around ritual centers that may have been originally founded on the sites of much smaller shrines and which may well have dated from even more ancient times. The clear suggestion is that these cities took both their alignments and ritual significance from shrines similar to the stupa that may have originally existed at those locales. Wheatley believed that many of these centers may date from periods that would be contemporaneous with ancient Egypt,[2] and the sites are thought by Fairservis to have served functions "similar to ritual centers of the Old Kingdom Egyptians and the Mayans." Wheatley writes, "Whenever . . . we trace back the characteristic urban form to its beginnings we arrive not at a settlement that is dominated by commercial relations, a primordial market, or at one that is focused on a citadel, an archetypical fortress, but rather at a ceremonial complex . . . even allowing for the biases in interpretation thus induced by the nature of the prescribed elements in the morphology of these complexes, the predominantly religious focus to the schedule of social activities associated with them leaves no room to doubt that we are dealing primarily with centers of ritual and [ceremony]."[3]

Also in China, we find the world's only surviving hieroglyphic language. This is the priestly language of the Na-khi-Dongba people (some say Naxi), which is referred to as the Dongba language in honor of the priests, themselves called *dongbas,* who are credited with having preserved it. So poorly understood is the Dongba language that popular references cannot even agree on a precise meaning for the term *dongba.* Some say that it means "wise person," some say "shaman," some say "sorcerer or wise man," some say "sage or wise person who can make divination and chant the scriptures," some say that it simply means "teacher." The written language, which, interestingly, is incapable of expressing every spoken thought in the Na-khi language, dealt primarily with issues of religion and cosmology, and is preserved in hundreds of thousands of ancient religious texts. Unfortunately, there are now only a handful of living dongbas who are still capable of reading the ancient texts.

Those familiar with the texts say that the ancient Dongba literature focused on a wide range of philosophical, cosmological, and religious subjects, including the formation of the universe and the world, and the processes of creation for all things, including humanity. The Dongba glyphs were apparently designed to express—first and foremost—concepts of creation and cosmology. An understanding of the Dongba language is considered by linguists to be of great value when researching the origins of classic Chinese language. The Dongba language also includes elements that are thought to reflect influences that likely came from India and Nepal.

The notion of an ancient Chinese/Tibetan hieroglyphic language that was associated with ritual shrines that were aligned like a stupa, that was preserved by tribal priests, and whose symbols and words focused primarily on concepts of cosmology runs closely parallel to any description we might reasonably devise for the Dogon. Certainly, in the context of a study of Dogon cosmology, it would make sense to invest time to pursue any apparent Na-khi-Dongba parallels to the Dogon.

Although, to my knowledge, there is no evidence to suggest any known historical contact specifically between the Dogon and the Dongba (Joseph Rock does claim that there were African influences on the Na-khi in his book *The Ancient Na-Khi Kingdom of Southwest China*), nonetheless the Dongba can be seen in many different ways to be effective Asian counterparts to the African Dogon. Both groups recount a long tradition as a nomadic priestly tribe, deeply invested in traditional cosmology and reverence for ancestors. Both preserve a highly stable culture that is based specifically on a cosmology that, to all outward appearances, sustained itself in substantially unchanged form for thousands of years. Both evolved a large body of drawn characters—in many cases, arguably the same characters—relating to the concepts of their cosmologies. Both credit honored ancestors/teachers with having introduced civilizing skills to the tribe. And both express their cosmology in distinctly similar terms, often with matching cosmological concepts and symbols.

Among Chinese scholars, there is some disagreement about the meaning of the name Na-khi. The first portion of the name, *na,* means "great," but can also mean "black" in the Na-khi language.[4] The Dogon word *na* and the Egyptian word *naa* can also mean "great or large," but neither means "black." Notwithstanding a traditional belief that the Na-khi may have originally been black Africans, this second meaning has been the source of some confusion because of an aspect of the cosmology that associates the concepts of good and evil with the colors white and black, respectively. Some researchers wonder why a tribe would knowingly name themselves using a term that means "evil." (Rock does claim that there were African influences on the Na-khi in *Ancient Na-Khi Kingdom of Southwest Africa,* and the word *Na-khi,* which Rock defines to mean "black man," is written using a glyph that is a stick figure of a man with a blackened face.)[5]

However, if we consider the word within the context of the cosmologies that we have studied, we know that the Dogon spider who correlates to the Egyptian mother goddess, Neith, carries the name of Nana and the title of Dada, which means "mother" in the Dogon language. We also recall interpreting the name Amazigh, a term that refers to the predynastic hunter tribes in Egypt, as combining the name of the Dogon hidden god Amma with the name of the Dogon Sigui festival (or the Egyptian word *skhai,* meaning "to celebrate a festival") to produce a name that we interpret to mean "celebrates or worships Amma." By this interpretation, the name would identify the tribal group in terms of the primary deity they worship. If we adopt this same approach to the interpretation of the name Na-khi, we might expect it to mean "celebrates Na" or "celebrates Neith," as we might also infer that the terms *Sigui, sigi,* and *skhai* could be taken as originating forms of the *xi* or *khi* suffixes of the interchangeable names Naxi and Na-khi. This view would make more sense if the word *khi* in the Dongba language could be shown to mean "to worship or to celebrate." As it turns out, some researchers interpret the name Na-khi to mean "worships all things black," a definition that upholds our supposition that the word *khi* can mean "to worship."

Also supportive of this interpretation is a definition in Rock's

A Na-Khi–English Encyclopedic Dictionary that indicates that the Dongba word *na* is a term that is used to refer to maternal relatives, so it could be properly interpreted to mean "mother."[6] From this perspective, the word *na* would have acquired the additional meaning of "black" in China precisely because the Na-khi had originally been black Africans, while the native Chinese residents were not.

An alternate Dongba figure used to represent the term *na* takes the form of a loop, similar to the looped string intersection (discussed in previous volumes of this series) that represents one way in which primordial threads are woven into matter.[7] This figure is closely associated in our parent cosmology with the mother goddess who is defined as the weaver of matter, the counterpart of the Dogon spider Nana or Dada.

Dongba spider glyph

The Na-khi people are thought to be the descendants of an ancient tribe called the Qiang, nomads who long ago migrated through the valleys in the northwest of China. Like the Dogon, who settled for a period near the Niger River, the Qiang eventually settled along the banks of the Jinsha River. Again like the Dogon, the Dongba now number about three hundred thousand individuals. They currently reside in a province in the northwest of China called Yunnan, an inhospitable area near the Tibetan Autonomous Region.

Like Dogon cosmology, the Dongba religion is reflected in all aspects of Na-khi culture and thereby contributes a foundational philosophy that is a guide to Dongba life. The Na-khi credit many of the civilizing aspects of their society to the instruction of a certain group of people, now worshipped as revered ancestors, who helped transform the Na-khi from wandering nomads into an organized society of farmers. These same instructors taught the Na-khi skills of metallurgy (iron and copper smelting) and are remembered for having made a great contribution to the social development of the tribe.

In *Sacred Symbols of the Dogon,* we drew parallels, based on Dogon descriptions of the egg-in-a-ball drawing, between the paired Dogon ancestors and the emergent deities of the Ennead and Ogdoad in ancient Egypt. In his 1957 book about the Na-khi-Dongba, *Forgotten Kingdom,* Peter Goullart tells us that the Na-khi religion is rooted in Buddhism and that they refer to the Tibetans as their elder brothers. Apparently unfamiliar with the innermost details of the Dogon and Buddhist cosmologies or with the concept of the egg-in-a-ball, Goullart expresses some confusion in regard to an ancient tribal belief regarding the mythical Na-khi ancestors. He writes, "Their ancestors are curiously linked with all the gods of the Indian pantheon and their claim [is] that the majority of their ancestors . . . came out of eggs. . . ."[8]

Dongba cosmology is traditionally seen as an outgrowth of the ancient Tibetan Bon religion, which conceived of the universe as having been formed from a primordial egg. Creation was associated with a great mother goddess whose symbol was a left-turning spiral. Among the pivotal processes of creation in the Dongba religion was the separation of earth and sky, an event that from one perspective gave birth to everything in the world. One important Na-khi creation myth, referred to as the Annals of Creation, is understood as a magical formula by which the power of the Word replays and subsequently recreates the foundations of the world.[9] The genealogy of the Dongba Mu ancestors (a term applied to the mythical founders of Dongba culture), like those of the Dogon, is said to be linked to the creation of a cosmic order. In Dongba myth, the universe is seen as the product of cosmic eggs of opposite nature and is created through a series of events that includes an incestuous marriage that pollutes the entire universe. It is this process that ultimately is responsible for the separation of earth and sky to form *Pu* and *Pa,* which are likely counterparts of the Dogon po and Egyptian pau. The concepts of Pu and Pa are symbolized by the Dongba female diviners and male priests, respectively.[10] The Dogon word for divination is *pey.*

Dongba culture and cosmology derives its physical expression largely in terms of a traditional ritual structure that is seen as a symbolic reflection of the cosmic and social forces that order their world. This structure, which is carefully aligned along east-west and north-south axes to the four cardinal points, is centered on a four-sided post called a *meedv,* which is thought of as the pillar of the universe. The radial symbolism of the structure involves both cosmology and biological kinship and reflects Na-khi concepts relating both to the emergence of cosmic space and to the proper social organization and division of labor within Na-khi society. The axes are associated in various ways with revered ancestors of the Na-khi and with what researchers refer to as primary oppositions, what we, in relation to Dogon cosmology, might call paired opposites.[11]

The Dongba glyph that symbolizes the concept of a barn full of grain takes the same familiar hemisphere shape that we associate with the Dogon granary and the Dogon concepts of mass or matter and with corresponding Buddhist concepts of essence or substance. According to Rock, this glyph carries the pronunciation of "o," which is also a Dongba term that is used as a name for god.[12] We may recall that the Dogon term *yu* refers to millet, a type of grain whose shape is compared to Amma and to the egg that resides within the egg-in-a-ball: .

The supreme deity of the Dongba religion, named O, is symbolized by a Dongba glyph, also pronounced "o," that takes a shape that is markedly similar to that of a stupa, with a circular base and a roughly conical or pyramidal shape rising to a flat roof:[13] .

Likewise, there is a secondary Dongba deity named O-gko-aw-gko who meditated on the creator/god O and thereby became a reality. We might correlate him to the Dogon mythical character Ogo (the likely counterpart to the Egyptian god of light, Aakhu), who plays the role of light in Dogon cosmology. According to the Dogon priests, Ogo saw Amma's creation and, thinking that he could create a world as perfect as Amma's, tore a square off Amma's placenta to create the reflection or image that we perceive as reality.

If aspects of Na-khi-Dongba cosmology seem distinctly familiar to us in terms of the cosmological model we've been studying, then so do the ancient concepts of written language as they have been preserved among the Dongba. The Dongba hieroglyphs grew up within the context of the ancient Tibetan and classic Chinese languages, to which they are said to be related. The Dongba hieroglyphs themselves are considered by linguists to be more ancient than the earliest surviving Shang dynasty oracle bone texts, and the glyphs of the language are compared with those of ancient Egypt and the Maya. One researcher reported that, at first glance, the Dongba script most closely resembled that of *The Egyptian Book of the Dead.* Such similarities would make sense, given the apparent relationship of the Dongba language to ancient Tibetan. In 1927, the Oxford University Press published the first English-language translation of the *Tibetan Book of the Dead,* so called by its compiler and editor, Walter Y. Evans-Wentz, because of the many parallels he found to *The Egyptian Book of the Dead.* A good general resource to everyday characters of the Dongba script is the *Naxi Dongba Pictographic Dictionary,* compiled by He Pingzheng and translated by Xuan Qin.

Of particular interest to this study is the fact that the Dongba glyphs are mnemonic in nature[14] and, like those in our Egyptian hieroglyph word examples, are read from left to right in columns and seem to reproduce meaning in the same way, simply by substituting concepts for glyphs to produce a symbolic sentence. Like our symbolic Egyptian hieroglyphic sentences, the reader is often left to infer transitional words, in much the same way that a traditional phonetic reading requires the reader to infer vowel sounds. A key difficulty to reading the script lies in understanding the proper concepts to be associated with each of the thousands of drawn shapes. My contention is that, for the Dongba as well as the Egyptians, these shapes and their corresponding meanings evolved first out of a common system of cosmology. It is for this reason that the Dongba language has been historically characterized as primarily expressing concepts of cosmology.

In further support of our initial view of similarities between Dongba and Egyptian hieroglyphic systems of writing, we notice that key Dongba glyphs take specific shapes and meanings that are already familiar to us based on our study of Dogon and Egyptian cosmology and language. For example, the Dongba character for day is the very image of the Dogon egg-in-a-ball drawing and virtually the same figure as the Egyptian sun glyph ☉, but complete with the bisecting lines or axes that, in the Dogon and Buddhist cosmologies, divide the figure into its four quarters. The presence of these lines strongly supports the notion that the figure derives from cosmology. We recall that one traditional explanation offered by Egyptologists for the shape of the Egyptian sun glyph is that the shape is an optical illusion that appears when a person stares too long at the sun, which causes the person to perceive a central dot. This explanation can explain the central dot of the Egyptian glyph, but not the bisecting lines of the Dongba character. Given the close relationship that is known to exist between Dongba writing and Dongba cosmology, it seems far more likely that the following glyph, like the other Dongba glyphs, derives from a distinctive and specific shape of a cosmology that the Dongba seem to share with the Dogon, the Buddhists, and the ancient Egyptians: (Dongba day glyph).

A similar character with attached rays descending below it is used by the Dongba to express the concept of sun. Like its Egyptian counterpart, the same Dongba glyph appears predictively in words whose meanings pertain either to periods of time or to times of day (such as *sunrise* and *sunset*). Likewise, it appears in words that bear an apparent geographic relationship to these times of day, such as *east* and *west,* the locales of sunrise and sunset, but significantly not in words such as *north* or *south,* which can be said to bear no relationship either to a period of time or to a time of day.

Likewise, the Dongba moon glyph takes a form that is distinctly similar to the Egyptian moon glyph , both without apparent logical rationale, since neither offers a precise match to the shape of the moon as it actually appears to us in the night sky: (Dongba moon glyph). Again, the presence of shared shape that is somewhat altered from the image any ancient culture might actually have seen in the sky supports an argument that the figure arises from a shared cosmology.

Another important cosmological glyph shape that takes similar form in both Egyptian and Dongba culture is that of the sieve , which for the Dogon serves as a conceptual metaphor for the processes by which particles are separated from waves during the formation of matter. The following is a representation of the Dongba sifter or sieve glyph: .

In Dongba cosmology, the world is the product of a process of creation that began with the separation of earth and sky. The Dongba glyph for this process, like the Dogon nummo fish drawing, can be seen as a composite of several significant drawings. The central portion of the Dongba glyph can be seen as a counterpart to the Dogon egg-in-a-ball, whose duality, in this case, is depicted using the traditional yin/yang figure. Below this egg is what in Dogon cosmology would be mass in its form as the primordial waves from which everything comes. For the Dongba, the concept of mass is referred to by its familiar cosmological key word *earth*. Above it, a wave draws up in tentlike form, as in Dogon cosmology, representing an initial expansion of space (the Dongba use the cosmological key word *sky*) and is encircled with a downward curl, as opposed to what is interpreted as an upward curl in the Dogon drawing, to create likely counterparts to the po pilu or Calabi-Yau space: . The dictionary caption for this glyph reads "Intercourse between the earth and sky: In Dongba religion, the earth and the sky gave birth to everything in the world."[15]

A second Dongba glyph that bears a likely relationship to the Dogon nummo fish is interpreted to represent a vagina and emphasizes the portion of the nummo fish that, in relation to the central womb, a vagina could properly reside: . (A matching Chinese symbol carries the meaning of "fish.") This glyph carries a pronunciation of "bi,"

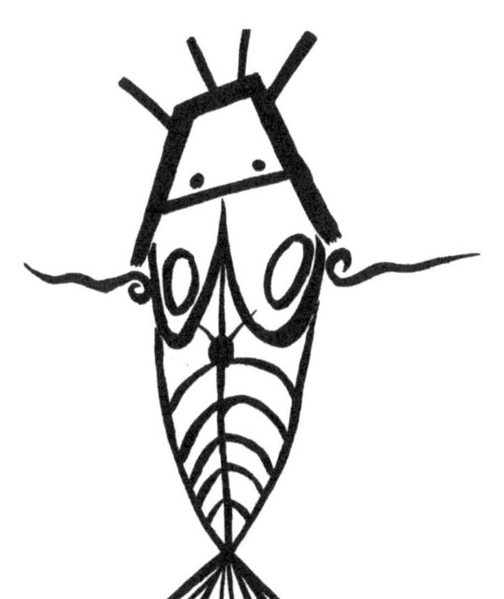

Dogon nummo fish drawing

the same as the Dongba sun glyph. From a cosmological standpoint, the shared pronunciation makes sense when we recall that the central feature of the Dogon nummo fish drawing is defined as the egg-in-a-ball, from which we believe the sun glyph takes its shape and meaning. However, we are also told by Rock that an alternate pronunciation for the Dongba vagina glyph is "ma." The same shape, when written next to the standing figure of a woman, means "mother" and is pronounced "a-ma."

Another Dongba word that carries apparent cosmological significance in its shape, pronunciation, and meaning is the word *ch wua ma la ma,* meaning "spider."[16] Looking to related words in Rock's dictionary, we realize that *ch wua* means "the number six." Likewise, the word *ch wua ma* refers to the sixth phase or stage of the moon and is written with the vagina glyph, which we take to be representative of the first stage of our cosmology. Based on these meanings and the general relationship of the spider mother Nana, or Dada, to Dogon cosmology, we might interpret the word *ch wua ma la ma* to mean "the sixth phase or stage of Amma," with the word *ama* again referring to the concept of mother.

In support of this view, we might note that the Dongba word for spider includes six distinct subfigures, including the left-most three-pronged branch (perhaps symbolic of the three branches of the tree of life), a central figure shaped like the Dogon po pilu or egg-of-the-world, and a left-turning spiral, the Dogon shape that characterizes the po pilu. The eight legs of the spider underscore the apparent choice of an insect to represent the eight vibrations of the po pilu:

Dongba spider glyph

Other Dongba glyphs might or might not support the suggestion of a relationship to Dogon cosmological shapes and meanings or to Egyptian glyph shapes. Although in most cases the *Naxi Dongba Pictographic Dictionary* does not discuss the specific symbolism of these shapes as they relate to Na-khi cosmology, Rock's *A Na-Khi–English Encyclopedia Dictionary* does include many cosmological words and makes occasional reference to cosmological meaning. My experience with these dictionaries has been that, in each referenced case, the specific details of that cosmology follow the anticipated pattern of our proposed parent cosmology.

It is clear, even based on the kind of superficial overview presented here, that the foundations of Tibetan and Chinese religion, society, and language rest on key principles of cosmology that fall into very close alignment with the Dogon, Egyptian, and Buddhist models upon which our studies have been predicated. My assertion is that the many diverse points of agreement between the parent cosmology and the Na-Khi cosmology are sufficient to allow us to effectively synchronize them with the Dongba language and, by inference, to the earlier Tibetan and later Chinese forms to which they so closely relate.

SIXTEEN

AS ABOVE, SO BELOW

The Chariot of Orion

The Hermetic phrase, "As above, so below," constitutes one of the signature phrases of ancient cosmology. It, along with the famous inscription found on an Egyptian temple to Neith that begins, "I am all that is, was, or will be," can be taken as characteristic statements of our proposed parent cosmology. The statement, "As above, so below," implies a cosmic relationship that astrophysicists have long suspected might be true, but have had no way of firmly corroborating. The first of these phrases suggests that events in the macrocosmic universe of stars and planets as we observe them are in some way fundamentally similar to events that occur in the microcosm, the universe of atoms, quarks, and Calabi-Yau spaces that fill the conceptual world beneath our plane of existence. We might even say that one of the as-yet-unclaimed prizes of modern astrophysics would be the discovery of a key to unlock those fundamental similarities and thereby effectively integrate what happens in the world *above us* with processes that transpire during the formation of matter, *below us.* As a cornerstone of the Hermetic system, the phrase, "As above, so below" is closely associated with many of the symbols of cosmology we have already discussed, beginning with the four primordial elements—water, fire, wind, and earth—so it can be seen to be intimately related to the same cosmologies we've been discussing. As such, the phrase effectively constitutes a statement of purpose for the cosmology.

Careful consideration of the many priestly statements documented by Griaule and Dieterlen in *The Pale Fox* reveals a perspective from which we can potentially interpret this phrase. To do so, we must first ask ourselves why the pyramid, which is perhaps the quintessential symbol of ancient cosmology, plays such a predominant role in ancient studies. In trying to answer this question, we see that the pyramids of Giza stand on the Giza plateau like a grand unsolved riddle. Partly because of their great age and partly because of their great size, they are in our face, posing an irresistible temptation to us in the same way that a single raised card might tempt us to choose it over all others during a child's card game of Old Maid. Generation after generation has puzzled over the intended meaning or purpose of these Herculean monuments, but somehow it seems that the more deeply we consider them, the more potential meaning we ultimately find in them.

Robert Bauval suggested that the three largest pyramids at Giza were arranged as they are to match the three belt stars of Orion, and he proposed in his Orion correlation theory that they were intended to record how these stars appeared from the earth on a particular date in antiquity. His widely researched evidence supported a convincing argument to explain what this arrangement of pyramids represented, but in my opinion failed to provide a compelling motive for why the ancients might have focused so much effort and attention on these particular stars.

Fortunately, as we have often found to be the case in the past, Dogon cosmology provides us with a sensible answer to the question why regarding the Giza pyramids. As it turns out, according to Griaule and Dieterlen, the Dogon also observe a tradition of placing large stones on a plateau as representations of key stars in their cosmology, and prominent among these are the three belt stars of Orion. The Dogon priests say that some of these symbolic stones are placed to represent stars as they appear in relation to one another in the heavens, and others, including the Giza stars, to represent the stars as they are seen from Earth. Griaule and Dieterlen write:

> In this series are the stars of which we have related the genesis and which to varying degrees will affect the life of man on the Fox's Earth. They will be represented by raised stones in places where the Dogon have hypothetically situated the episodes of the first sixty-six years of man's life on Earth, both for purposes of initiation and for the performance of certain rites. In particular, the majority of those heavenly bodies . . . are . . . represented by raised stones on the rocky plateau of *ka donnolo,* "plateau of *ka. . . .*" One says that [some] stars . . . are placed as they are in the sky; those of the *ka donnolo* are placed as people have seen them in the world (which is to say from the Earth).[1]

Like virtually all important words of Dogon cosmology, Griaule and Dieterlen tell us that the Dogon term for the belt stars, *tolo atanu,* carries a second, logically disconnected meaning, in this case, "deputies." We can confirm a likely relationship between the Dogon and Egyptian concepts by looking to the Egyptian word for deputy, which Budge lists as *atenu.*[2] The Dogon priests say that the purpose of placing the stones that represent the three belt stars is to call attention to something they refer to as the chariot of Orion. Griaule and Dieterlen write, "Thus in stellar space, the Chariot is the symbol of Amma's seat; it surrounds *atanu,* the Belt, otherwise known as the three 'deputies.'"[3]

In 2008, in pursuit of a deeper understanding of what he deemed to be an original architectural plan on the Giza plateau, an acquaintance of mine and fellow author and researcher of ancient mysteries named Scott Creighton applied careful geometric analysis to a photograph of the three belt stars of Orion, then overlaid that same geometric pattern on an aerial photograph of the Giza pyramids, reproduced at the same scale. With these two photographs, Scott demonstrated a very close match. I commented to him that, if the scale and geometry of the two photographs truly do match, then we should be able to demonstrate a relationship based on the distance between any two Giza pyramids (measured in cubits) and the distance between the corresponding Orion belt stars they represent (measured in light years).

Further consideration of my own comment induced me to explore ancient Egyptian words for cubit to see what information might be gleaned from a symbolic reading of the words. What I discovered was an Egyptian term for cubit, *aakhu meh,* that is, in fact, a compound word in the familiar English-language sense, consisting of two other Egyptian words. The first word, *aakhu,* is both the Egyptian word for light and the name of the Egyptian god of light. The second word, *meh,* is a root that we take to mean measure. Together, I interpret these words to read "light measure," the very concept I had felt might be correlated in some regard to cubits based on Creighton's belt-star geometry.

Realizing that the length of an Egyptian cubit is a value that was inferred by modern researchers based on the dimensions of the Great Pyramid at Giza, I noted that the pyramid measures 440 cubits per side and 280 cubits in height. I put two and two together and decided to explore whether those numbers might bear a known relationship to the belt stars of Orion, the stellar bodies thought by Bauval to be represented by the pyramids. Using an online search engine on my computer, I keyed the numbers 440 and 280 along with the word *Orion* and was pleasantly surprised to turn up references to an obscure, virtually invisible cosmic structure called Barnard's Loop, a structure that I had previously not known to even exist.

In accordance with the traditional symbolism of the Giza pyramids,

Barnard's Loop is part of the Orion nebula, an area in space that is thought to have been created by a supernova (the death of a star) and a region where new stars are reborn. Barnard's Loop measures 440 light years by 280 light years, the very same dimensions in light years as the cubit measurements of the Great Pyramid. The light that is emitted by Barnard's Loop is so very weak that it can only be imaged through time-lapse photography (some researchers suggest that the loop can be faintly seen in a very blackened night sky). When it is successfully imaged, its appearance takes the familiar shape of a spiraling coil ୧, one that encircles the belt stars of Orion.

The reader may recall earlier discussions of the po pilu, the tiny

Barnard's Loop surrounds Orion's Belt.
Edited photo by Rich Richins, www.enchanted skies.net; www.enchantedskies.net/BarnardsLoopMap.tif.

Dogon egg of the world that houses the seven vibrations of the threads of matter and that we take as a counterpart to the wrapped up dimensions of the Calabi-Yau space in string theory or torsion theory. The Dogon priests say that this enclosure is created when wind (taken as a metaphor for the concept of vibration) causes mass to become circular and form a tiny egg. The seven vibrations within this egg are interpreted as seven rays of a star of increasing length and are characterized by the spiral that can be drawn to inscribe the endpoints of those rays. The seventh ray is said to eventually grow long enough to pierce or burst the egg. My research regarding Barnard's Loop revealed that it constitutes a stellar bubble, a structure that is formed when the stellar wind causes matter to become circular and form a much larger kind of egg. This egg, like the po pilu, is also calculated to eventually burst.

The reader may also recall the shapes by which the Dogon priests represent each of the seven vibrations of the po pilu. The second of these, which represents the second of the seven phases of vibration, is a close match for a modern-day photograph of a supernova, the event that is credited by modern astronomers with the formation of Barnard's Loop.

When the Dogon speak of the belt stars of Orion as deputies, a word that for the Egyptians means "to rule in place of another," the unspoken implication is that they consider the stars to be deputies of the two binary stars of Sirius, the sunlike star that they call *sigi tolo*

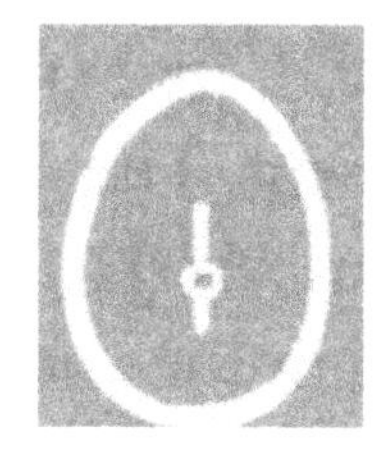

Second vibration of the po pilu

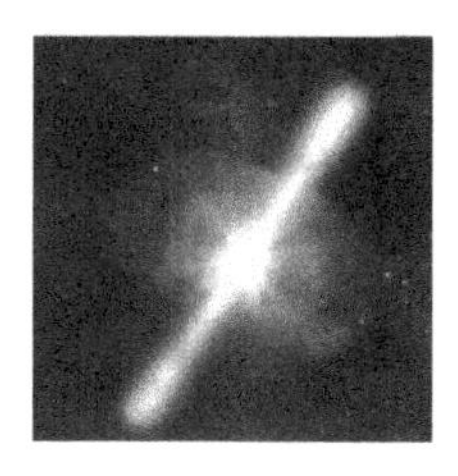

Photograph of a supernova

(star of the Sigui festival or fifty-year star, in reference to the orbital period of the two stars) and the dense dwarf star *po tolo* (star of deep beginning). Scientists also consider this dwarf star to be the remnant of a supernova, perhaps the very same event that is credited with having formed Barnard's Loop.

The motion of these two stars—one around the other—re-creates the figure of the Dogon egg-in-a-ball, while Barnard's Loop, which encircles the deputies of Sirius in the shape of a spiral, recreates the same figure ౿ that is used to characterize the po pilu. All together, this area of space is referred to by the Dogon priests as the star of the po. Regarding this astronomic region, Griaule and Dieterlen write:

> The star of the fonio [fonio is the smallest grain, a reference to the dwarf star of Sirius], *po tolo* turns around Sirius, *sigi tolo*. The revolution takes fifty years. It is the most important of all the stars and plays a key role—in all the spiraling star worlds Amma formed; as it is considered to be the center of the stellar world. . . . The star of the po, a double of the germ of the whole creation, will also be of considerable value to human beings. Later, when they see it in the sky, it will be testimony of the renewal of the world for them; the image of the "womb" of Amma, who preserves the basic signs for the whole creation within himself and who keeps the *po pilu,* which that star represents in the sky. Therefore one says that it is like "the egg of the world," *"aduno talu"* and like "Amma's eye," *amma giri,* "the guide of the universe."[4]

When we step back from the figure of Orion given above as part of the image of Barnard's Loop, we see that Barnard's Loop takes on the appearance of the wheel of a chariot and gives the impression that Orion, the hunter, is actually standing in a chariot. It is to this resemblance that we credit the Dogon symbolic term for Barnard's Loop as the chariot of Orion. It is important to note that when modern-day Dogon priests diagram the concept of the chariot of Orion, they do so, at least publicly, in relation to the visible stars in the constellation of Orion, which simply form a kind of parallelogram that surrounds the belt stars but bear no discernable resemblance either to a chariot or to the po pilu, as the Dogon priests insist they should. My contention is that the true symbolism of the term is to Barnard's Loop, an effectively invisible structure that is a near-exact conceptual counterpart to the po pilu and that the Dogon priests suggest will only be seen "later."

Based on this interpretation, Barnard's Loop is the structure represented by the Dogon chariot of Orion and was considered by the Dogon priests to be of very great importance to humanity because it demonstrates two critical points relating to cosmological creation. First, it suggests that the reality we perceive, the Dogon Third World, may not be, as we surmise it to be, the final end-product of creation, but rather simply another routine stop along a much grander path of creation, one that continually repeats or renews itself from world to world. Second, it provides us with a working example, taken from within the domain of our own reality and, perhaps more importantly, from our own local neighborhood in the universe, of the egg of the world, the Dogon po pilu, or string theory's Calabi-Yau space. The suggestion is that we may be able to draw inferences based on this cosmic structure regarding the likely nature of corresponding microscopic structures. The potential benefit to humanity of that one critical example could be incalculable.

Support for this view is found, again, in the Egyptian hieroglyphic language. In *Sacred Symbols of the Dogon,* we discussed correlations between the Egyptian concept of a chamber, hall, or pylon and the wrapped-up dimensions of a Calabi-Yau space, a likely counterpart to the Egyptian underworld or "other world." These center on what we refer to as the chamber or pylon glyph, which carries the pronunciation "urit."[5] Budge defines three significant meanings for words pronounced "urit" or "urr-t": one he defines to mean "pylon or chamber," one he terms a name for the Egyptian underworld, and one he defines as meaning "chariot."[6]

If we pursue the suggestion that in our parent cosmology the concept of above refers specifically to the spiral of Barnard's Loop, we find we can support this notion simply by taking notice of the form of an Egyptian hieroglyphic word *au,* meaning "above." Following the same interpretive method that we have applied to many other Egyptian words, we find that it reads:

ABOVE (AU)

Above		That which is the spiral or loop
		(see Budge, p. 30b)

Notwithstanding supportive Egyptian word references, the concept of the chariot of Orion is one that raises an interesting question regarding the origins of our parent cosmology. In the view of most traditional Egyptologists, images of chariots do not appear in the Egyptian archeological record until around 1800 BC, while the evidence I have cited in our studies suggests that the cosmology must have been first presented in Egypt sometime prior to 3000 BC and that one of its likely original purposes—to provide humanity with a concrete link between the macrocosm and the microcosm—seems inextricably intertwined with the concept of the chariot of Orion. If the chariot was unknown in Egypt until more than 1,200 years later, how could this be?

One potentially credible answer to this question could rest on a premise that the cosmology was first implemented in India or Asia, two regions in which the chariot is known to have existed by 3000 BC and where its use as a symbol can be understood as reasonable. Another lies with a little-known tradition that is associated with the Nazca lines in Peru, dating from around 3300 BC. This refers to a concept called the wheel of Orion, which, like the chariot of Orion, was said to center on the three belt stars and to mark a cosmic birthplace of stars. From this perspective, the original symbolism would have been to the concept of a wheel, and only later supplanted by references to a chariot.

The notion of Barnard's Loop as an astronomic structure in the form of either a wheel or a chariot and as being associated with light measures, dying stars, and the birthplace of suns is one that is also specifically upheld in the Vedic tradition. According to the India Profile website:

> In Vedic imagery, Surya or the Vedic god of light is pictured as riding a golden chariot with just one wheel, driven by several powerful steeds that carry him at the speed of three hundred and sixty-four leagues per wink! Riding through the sky he keeps a watchful eye on the world. Who can match the brilliance of the sun! [M]editate upon the celestial object for but a second and it rises to create everlasting wonder. Who created the sun on which all life is dependent? Is the big dying star just a scientific phenomenon or is it the powerful and majestic Lord Surya . . . ?[7]

In the Hindu tradition, the term Surya, the name of the god deemed to have been the creator of our sun, is often, in my view, misapplied to the sun itself. However, the suggestion is that the original symbolism was to this same chariot, Barnard's Loop. It was considered to have been the most readily apparent image of divinity available to man. Esoterically, it represented the point where the manifest and unmanifest worlds meet or unite, the very definition of the Dogon po pilu.

From the perspective of astronomers, Barnard's Loop was the product of a supernova. The Dogon associate Barnard's Loop with the stars of Sirius, one of which, the dwarf star Sirius B, is also thought by astronomers to be the remnant of a supernova. For the Dogon, these stars constitute a local cosmic center, by whose influence the stars in our region of the universe are understood to somehow maintain their proper positions.

The aligned ritual structure, the stupa or granary, which we said represents the conceptual starting point or intellectual bottom of the cosmology for an initiate, revolves around the central point and the six axis rays by which it is oriented to the cardinal points of the universe. These can be seen as counterparts to the seven rays of the po pilu. Ultimately, they bring us full circle to Barnard's Loop, a macrocosmic egg defined as a place of the rebirth of stars, formed in the same way and with characteristics that largely match the po pilu, and the structure that is deemed to define and sustain the orientation of our region of the universe. Appropriately, this represents the conceptual top or ending-point of the cosmology.

My belief is that it was a desire to communicate this critical clue regarding the true nature of creation—the existence of a structure within our own range of view that illustrates one of the key processes of existence—that justified the immense effort that must have gone into building the Giza pyramids. From this perspective, it would be much as Bauval and others have surmised, that the Giza pyramids do represent the belt stars of Orion as viewed from the earth and that the Great Pyramid does indeed encode numeric values that will be important to our understanding of creation. From this perspective, it would have been a desire to transmit these very same facts and clues that lay at the heart of the plan of the ancient cosmologies.

SEVENTEEN

RECAPPING THE PLAN OF THE COSMOLOGY

Having outlined the likely conceptual boundaries of our proposed ancient parent cosmology, now might be a very good moment for us to stop and recap what we have learned from our explorations. Our study of ancient cosmology has been an intellectual journey that has taken us from the egg-in-a-ball and the massless waves of the Dogon First World of matter to the formation of the po pilu, or egg of the world, in the Dogon Second World of matter, and to the po, or atom, which is the first finished creation of our own world, the Dogon Third World of matter. We then came around full circle to the nearly imperceptible chariot of Orion, which is a stellar bubble, and to the likely macrocosmic counterpart to the po pilu, a structure whose very existence invites us to meditate on long-past classic notions of worlds within worlds, ages within ages, and cycles within cycles.

We believe, based on the explicit statements of many different ancient cultures, that ancient cosmology was intended as a civilizing world plan for humanity. Our primary reference to that plan comes out of Dogon cosmology, which we believe to be the most substantially complete extant version, one that is greatly enhanced by the ability of living, knowledgeable Dogon priests to explicate its innermost meanings and subtle intentions. We have come to know these primarily through *The Pale Fox* and other significant writings that reflect the long, careful studies of Griaule and Dieterlen.

The completeness and accuracy of the Dogon references are independently affirmed by Buddhist stupa symbolism—almost wholly without regard to Dogon cosmology—by Snodgrass in his *Symbolism of the Stupa.* Buddhist cosmology, as described by Snodgrass, touches on virtually every key Dogon theme, symbol, and concept. Likewise, Snodgrass attributes the correct transmission of this cosmology for the benefit of each new generation to the symbolic attributes of the stupa, an aligned ritual shrine similar to the Dogon granary. Each serves as a kind of grand mnemonic for its respective cosmology. Further substantiation for the Dogon/Buddhist cosmological system can be drawn from ancient Egyptian cosmology, whose often fragmented but well-documented symbols and words have been shown to specifically mirror and uphold key relationships defined within the cosmology.

Central to our Dogon/Egyptian comparisons are the ongoing, demonstrable parallels between Dogon and Egyptian cosmological words. These comparisons would not have been possible without—and have been incalculably aided by—the many word definitions drawn from Calame-Griaule's *Dictionnaire Dogon.* Likewise, they depend on word entries in Budge's *An Egyptian Hieroglyphic Dictionary,* which offers a view of Egyptian word pronunciations and meanings that have gone out of favor with traditional Egyptologists since its publication but that nonetheless bear a consistent and close resemblance to Dogon and African cosmological terms, meanings, and pronunciations. My contention is that Budge's dictionary should not be capable of sustaining these ongoing resemblances if, in fact, it grossly misrepresents Egyptian word meanings and pronunciations, as some traditional Egyptologists suggest.

The apparent purpose of this ancient plan of cosmology was to teach civilizing skills to mankind, organized and presented in such a way as to help orient us to key concepts of creational science, while at the same time providing us with an ongoing creation tradition. Likewise, the plan of the cosmology provided a specific system for civic organization, with careful divisions of labor designed to foster the kind of sustainable economy that would be required by a civilized society. The following are among the civilizing skills that we can attribute to this plan:

1. **Skills related to the construction of the aligned ritual stupa or granary.** These include the establishment of practical units of measure relating to body parts, specifically the cubit, based on the forearm, the palm, and the digit or finger. They also include the introduction of basic skills of math and geometry, including the ability to plot points and draw a line, arc, and circle. The plan began with the Dogon egg-in-a-ball, which represents the point of perception of a massless wave and the point of conception of a fertilized egg. This base plan constituted both a working sundial and a stellar observatory, which facilitated the establishment of functional units of time, including the sixty-minute hour, twenty-four-hour day, ten-day week, thirty-day month (three weeks), twelve-week seasons (four months each), and 360-day year, all of which are factors of the grand precessional cycle of 28,500 years. Stages of the stupa/granary plan were correlated symbolically to categories of signs and to stages in the formation of a Word. Among the skills of construction that were taught was a method by which to organize the plan of construction that included four stages: bummo, yala, tonu, and toymu.

2. **The introduction of the concepts of fibers and threads, then those of spinning and weaving.** This set of civilizing skills, like the next stage of matter to which they seemingly correlate, begins with the concept of threads as underlying components of matter and ends with the notion that these threads are ultimately woven to form matter. The wavelike fibers were compared to water as it evaporates in the sun. The art of weaving is equated with the act of speech, and the idea is introduced that words are woven into the cloth, much as DNA can be said to be woven into each living cell.
3. **Instruction in rhythm and music, beginning with the construction of a drum.** This civilizing skill can be seen to reinforce the cosmological notion that matter is the product of vibrations.
4. **Instruction in cultivation of the land and agriculture.** This civilizing skill can be seen to reinforce the cosmological notion that vibrating threads create mass, known by its cosmological key word *earth*. Cultivation of the land was correlated to the art of weaving, so initiates were taught to plow a field in the same manner in which cloth is woven. Clearing a plot of ground was compared to the act of weaving a cloth. Likewise, cultivation was compared to an act of speech.
5. **Introduction of the concept of clothing.** Once threads had been woven into cloth, then cloth could be turned into clothing. This civilizing skill can be seen to reinforce the cosmological notion that threads form membranes. The ongoing metaphor of a relationship between the arts of weaving and speech is reinforced by the statement that to be naked is to be speechless.
6. **Instruction in the skills of metallurgy.** A forge was established on the flat roof of the granary, and instruction was given in how to form tools from copper and iron. The smith was

charged with fashioning the implements of agriculture, but as part of a carefully systematized division of labor, was forbidden to use them with his own hand.

7. **Introduction of the art of pottery.** The Dogon say that the art of pottery was discovered by accident when the wife of the smith left a clay pot that was to dry in the sun a little too close to the hot forge and unintentionally fired the first clay pot. This skill can be seen to reinforce, through the metaphoric image of clay pots that hold water, the cosmological notion that waves of matter somehow collect to form particles. Egyptian word references suggest that the art of baking bread, the next logical civilizing stage beyond the cultivation and harvesting of grains, was also taught at this stage.
8. **Conceptualization of the stupa/granary as a world system.** The completed granary was further defined as a grand metaphor for the organization of the world. Its four faces were associated with the star groups whose risings and settings regulated the agricultural cycle. Its eight interior chambers were used to categorize and store the eight cultivated grains. Its four staircases were associated with a system of botany and zoology, so they represented major classes of plants and animals, while each step represented a family or order of plants or animals. Concepts and symbols that relate to the finished stupa or granary seem to express themselves, not as a part of the base parent cosmology, but as a later outgrowth of that cosmology in the form of astrology and the classic signs of the zodiac. In this way, the aligned ritual structure became the symbolic repository of an entire world system of instructed knowledge.
9. **The importance of symbolism, ritual, and sacrifice.** Prior to the eventual introduction of written language, symbols and spoken cosmological key words constituted an effective

mnemonic language of the cosmology. As we have shown, these symbols were defined first in relation to component stages of creation. Rituals were introduced initially to mark the risings and settings of significant stars, particularly those important to the regulation of agriculture. The likely eventual goal of these rituals may have been to highlight the presence of macrocosmic structures important to cosmological science, such as Barnard's Loop. The ritual placement of stones on a plateau to represent stars can be seen to have grown into a tradition of pyramid building—based initially on principles that define the alignment and construction of a stupa—in Egypt and elsewhere. The concept of ritual sacrifice as Griaule understood it emphasized the importance of cycles in the perpetuation of life, culture, and cosmology. An act of ritual sacrifice represented a return of life force to the earth as the completion of a cycle through the spilling of blood, which was absorbed into the earth. This may have been meant to underscore the importance of sustaining an ecological balance in the world.

10. **The introduction of written language.** It was an apparent part of the original plan of the parent cosmology to ultimately introduce the concept of written language among the final civilizing skills. This inference is supported both by emphasis within the cosmology on processes that culminate in the formation of a Word from signs and by the likely preassignment of phonetic concepts—the same phonetic values later assigned to Egyptian glyphs—that appear to govern the formation of spoken Dogon and Egyptian cosmological words (see the discussions of Egyptian phonetic values in *Sacred Symbols of the Dogon*). Written language would then have been the logical extension of these processes and could, understandably, have been initially based on the well-defined shapes/concepts

> that had already been evoked by the cosmology. These may have been used initially as they are in the Dongba language to primarily express concepts of creation and cosmology.

Although the Dogon and Buddhist cosmologies represent esoteric systems, they are clearly systems in which any properly motivated person is ultimately allowed to learn the most intimate and innermost details. As such, this can hardly be seen to constitute a secret that was meant to be kept from humanity. But in that case, from whom was the secret ultimately intended to be kept? The reasonable suggestion is that the system was designed to conceal aspects of the cosmology from some third party. There are clues to be found in the various world cosmologies we have studied, that the theoretical authors and organizers of our parent cosmology may have been delegated by some larger group to bring civilizing skills to mankind. The serpent is a recurring symbol of these teachers, and in some traditions, like Christianity, the serpent is held responsible for a grave transgression, one that seemingly involved the unauthorized transmission of knowledge to early humanity. Perhaps these ancestors/teachers had been dispatched to instruct mankind in each of the basic civilizing skills, however with a specific prohibition against instructing us in the innermost details of science.

However, it is easy to imagine that, like any conscientious parental figure, these teachers came to realize that the somewhat simplified explanations that we sometimes find ourselves obligated to give our children should never deliberately mislead or misinform them, rather, even in simplified form, they should somehow still reflect an ultimate truth. It would have made no sense whatsoever to provide mankind with a creation tradition that did not reflect actual events as they transpire in nature. So it appears that an idealized plan was eventually formulated for the cosmology, one that, in the end, ran directly parallel to and carefully reflected many of the most intimate details of creation science. (This conclusion would be consistent with the apparent choice within this same instructional mind-set to provide us with units of time that were factors of the precession cycle and units of measure that were based on actual dimensions of the earth.)

From this perspective, the esoteric tradition would have been devised to hide any obvious relationships between the cosmology and deep science from that original delegating body. Under the rules of the esoteric tradition, the scientific aspects of the cosmology, which correspond to the innermost secrets of the tradition, would be affirmed only to longtime students or initiates and would be functionally disavowed to any other casual questioner. So even in deepest antiquity, among the most revered ancient teachers of humanity, that is, the serpents who were originally almost uniformly revered and honored by mankind and who were often associated with the stars of Sirius, we find familiar evidence of all-too-human failings such as insubordination and calculated deception. Such acts were apparently justifiable when undertaken on our behalf as an apparently unavoidable consequence of the plan of the ancient cosmologies.

If there is one common thread that runs through the tapestry of ancient cosmology as we have come to know it through the volumes of this series, it is a clear sense that many of its elements appear to be the product of a deliberative, instructive hand. Likewise, we have shown that if we go beyond a superficial study of ancient cosmology in any given culture, what we ultimately uncover is an explicit memory of cosmology as an organized instructional form—one that each culture associates with the ancient acquisition of civilizing skills. The many details of this cosmology—couched as it was in its complex parallel themes and defined by a single shared set of well-conceived symbols—

are reflective of the very great capabilities of whatever ancient authority may have originally organized and presented it on our behalf. It is within this tradition that we find the roots of modern religion, astronomy, mathematics, and creational science, a source from which the many rich mythologies of the ancient world may well have sprung, as well as a credible ancient antecedent to—or foundation for—the earliest forms of symbolic writing. It is by way of an understanding of this cosmology that we can perhaps best grasp the inexplicable commonality in form and meaning of mythological ancient symbols that has been demonstrated worldwide—not as an outgrowth of parallel development or inbred psychology, but rather as a precious artifact of mutual instruction by a commonly revered set of teachers. Ultimately, it is to the very careful plan of this instructed legacy that we may most sensibly attribute the cosmological origin of myth and symbol.

NOTES

Introduction

1. Griaule, *Conversations with Ogotemmeli,* 18.
2. Ibid., xv.
3. Sagan, *Broca's Brain,* chap. 6.
4. Van Beek, *Dogon: Africa's People of the Cliffs,* 103.
5. Snodgrass, *Symbolism of the Stupa,* v.
6. Scranton, "Revisiting Griaule's Dogon cosmology," 24.
7. Griaule and Dieterlen, *Pale Fox,* 20–24.
8. Scranton, *Sacred Symbols of the Dogon,* references throughout.

Chapter 1. Concepts of Comparative Cosmology

1. Griaule and Dieterlen, *Pale Fox,* 240.
2. Snodgrass, *Symbolism of the Stupa,* 117–18.
3. Griaule and Dieterlen, *Pale Fox,* 82.
4. Ibid., 81.
5. Budge, *Egyptian Hieroglyphic Dictionary,* 54a.

Chapter 2. Signature Signs of the Parent Cosmology

1. Calame-Griaule, *Dictionnaire Dogon,* 70.
2. Budge, *Egyptian Hieroglyphic Dictionary,* 23ab.
3. Griaule and Dieterlen, *Pale Fox,* 131.
4. Budge, *Egyptian Hieroglyphic Dictionary,* 230b–231a.
5. Long, *Alpha: Myths of Creation,* 65.

6. Best, *Maori Religion and Mythology,* 33.
7. Griaule, *Conversations with Ogotemmeli,* 77.
8. Snodgrass, *Symbolism of the Stupa,* 14–17.
9. Griaule, *Conversations with Ogotemmeli,* 32.

Chapter 3. The Dogon Mythological Structure of Matter

1. Griaule and Dieterlen, *Pale Fox,* 70.
2. Scranton, *Science of the Dogon,* 54–73.
3. Griaule and Dieterlen, *Pale Fox,* 110.
4. Ibid., 131.
5. Ibid., 133–35.
6. Budge, *Egyptian Hieroglyphic Dictionary,* cxxvi.
7. Ibid., 34a.
8. Ibid., 34b.
9. Ibid., 604a.
10. Ibid., 606a.
11. Griaule and Dieterlen, *Pale Fox,* 554.
12. Ibid., 137.
13. Ibid., 138.
14. Ibid., 417.
15. Budge, *Egyptian Hieroglyphic Dictionary,* 349b.
16. Ibid., cxxvi.
17. Griaule and Dieterlen, *Pale Fox,* 267.
18. Calame-Griaule, *Dictionnaire Dogon,* 12.

Chapter 4. Symbolism

1. Budge, *Egyptian Hieroglyphic Dictionary,* lxv.
2. Calame-Griaule, *Dictionnaire Dogon,* 3.
3. Budge, *Egyptian Hieroglyphic Dictionary,* 23a.
4. Ibid., 120a.
5. Ibid., 500a.
6. Ibid., 628b.
7. Ibid., 69a.
8. Griaule and Dieterlen, *Pale Fox,* 59.
9. Ibid., 58–59.

10. Snodgrass, *Symbolism of the Stupa,* 9.
11. Swanson, "The Concept of Change in the Great Treatise," in H. Rosemont Jr., ed., *Explorations in Early Chinese Cosmology,* 75.
12. Budge, *Egyptian Hieroglyphic Dictionary,* 743b.
13. Ibid., 746b.
14. Scranton, *Sacred Symbols of the Dogon,* chap. 6.

Chapter 5. Guiding Metaphors of the Cosmology

1. Griaule and Dieterlen, *Pale Fox,* 95.
2. Budge, *Egyptian Hieroglyphic Dictionary,* 214a.
3. Ibid., 133b.
4. Ibid., 838a.
5. Ibid., 834a.
6. Ibid., 161b.
7. Ibid., 145a.
8. Ibid., 562a.
9. Calame-Griaule, *Dictionnaire Dogon,* 315–16.
10. Griaule and Dieterlen, *Pale Fox,* 136–41.

Chapter 6. The Egg-In-a-Ball

1. Griaule and Dieterlen, *Pale Fox,* 130–31.
2. Ibid., 83.
3. Budge, *Egyptian Hieroglyphic Dictionary,* 44a.
4. Ibid., 266b.
5. Griaule and Dieterlen, *Pale Fox,* 82.
6. Budge, *Egyptian Hieroglyphic Dictionary,* 50b.
7. Ibid., 899b.
8. Griaule and Dieterlen, *Pale Fox,* 85.
9. Budge, *Egyptian Hieroglyphic Dictionary,* 45b.
10. Griaule and Dieterlen, *Pale Fox,* 86–92.
11. Griaule, *Conversations with Ogotemmeli,* 24–25.

Chapter 7. The Aligned Ritual Shrine

1. Snodgrass, *Symbolism of the Stupa,* 17.
2. Ibid., vii.
3. Budge, *Egyptian Hieroglyphic Dictionary,* 614a.

4. Ibid., cvii.
5. Scranton, *Sacred Symbols of the Dogon,* 136–37.
6. Budge, *Egyptian Hieroglyphic Dictionary,* 616a.
7. Snodgrass, *Symbolism of the Stupa,* 1.
8. Ibid., 24.
9. Ibid., 209.
10. Griaule and Dieterlen, *Pale Fox,* 64.
11. Snodgrass, *Symbolism of the Stupa,* 215.
12. Griaule, *Conversations with Ogotemmeli,* 33.
13. Scranton, *Science of the Dogon,* 131.
14. Griaule, *Conversations with Ogotemmeli,* 33.
15. Ibid., 37.
16. Ibid., 51.
17. Budge, *Egyptian Hieroglyphic Dictionary,* 633b.

Chapter 8. The Elemental Deities

1. Hart, *Dictionary of Egyptian Gods and Goddesses,* 65.
2. Ibid., 47.
3. Lucas, *Religion of the Yorubas,* 21.
4. Winters, IPOAA Magazine, "The Proto-Saharan Religions," www.ipoaa.com/proto_saharan_religions.htm (accessed April 27, 2010).
5. Griaule and Dieterlen, *Pale Fox,* 82.
6. Ibid., 81.
7. Budge, *Egyptian Hieroglyphic Dictionary,* 53a.
8. Ibid., 51b.
9. Ibid., 546b.
10. Ibid., 548a.
11. Ibid., 53a.
12. Ibid., 25b.
13. Ibid., 4a, 31b.
14. Ibid., 256b.
15. Ibid., 255b.
16. Ibid., 570b.
17. Ibid., 570a.
18. Ibid., 250a.
19. Ibid., 250a.

20. Ibid., 231a.
21. Ibid., 230b.
22. Ibid., 231a.
23. Snodgrass, *Symbolism of the Stupa,* 1.
24. Budge, *Egyptian Hieroglyphic Dictionary,* 830a.
25. Ibid., 828b.
26. Ibid., 830b.
27. Ibid., 909a.
28. Ibid., 233b.
29. Calame-Griaule, *Dictionnaire Dogon,* 215.
30. Budge, *Egyptian Hieroglyphic Dictionary,* 23a.
31. Ibid.

Chapter 9. The Concept of the Primordial Egg

1. Budge, *Egyptian Hieroglyphic Dictionary,* 679b–80a.
2. Ibid., 680a.

Chapter 10. The Concept of the Divine Word

1. Griaule and Dieterlen, *Pale Fox,* 58.
2. Ibid., 59.
3. Ibid., 417.
4. Budge, *Egyptian Hieroglyphic Dictionary,* 240a.
5. Ibid., 913a.
6. Scranton, *Sacred Symbols of the Dogon,* 84.
7. Budge, *Egyptian Hieroglyphic Dictionary,* 332b.
8. Ibid., 655a.
9. Ibid., 54a.

Chapter 11. The Concept of the Fish

1. Griaule, *Conversations with Ogotemmeli,* 30–35.
2. BBC News, "Fish Fossil Clue to Origin of Sex," http://news.bbc.co.uk/2/hi/science/nature/7909984.stm (accessed April 27, 2010).
3. Scranton, *Sacred Symbols of the Dogon,* chap. 8.
4. Budge, *Egyptian Hieroglyphic Dictionary,* 12b.

Chapter 12. Deities

1. Budge, *Egyptian Hieroglyphic Dictionary,* 349b.
2. Ibid., 401a.

Chapter 13. Civilizing Skills

1. Budge, *Egyptian Hieroglyphic Dictionary,* 751b.
2. Ibid.
3. Griaule, *Conversations with Ogotemmeli,* 16–230.
4. Budge, *Egyptian Hieroglyphic Dictionary,* 753b.
5. Ibid., 736a.
6. Griaule, *Conversations with Ogotemmeli,* 20.
7. Budge, *Egyptian Hieroglyphic Dictionary,* 751a.
8. Ibid., 754b.
9. Ibid., 751b.
10. Ibid., 750b.
11. Ibid., 754a.
12. Ibid., 751b.
13. Ibid., 754b.
14. Ibid., 750b.
15. Ibid., 751b.
16. Griaule, *Conversations with Ogotemmeli,* 158–59.
17. Budge, *Egyptian Hieroglyphic Dictionary,* 754a.

Chapter 14. Written Language

1. Scranton, *Sacred Symbols of the Dogon,* 69.
2. Griaule and Dieterlen, *Pale Fox,* 33.
3. Scranton, *Sacred Symbols of the Dogon,* chap. 11.
4. Scranton, *Sacred Symbols of the Dogon,* 110; Budge, *Egyptian Hieroglyphic Dictionary,* 190a.
5. Renouf, *Egyptian Book of the Dead,* 187.
6. Budge, *Egyptian Hieroglyphic Dictionary,* 180b.
7. Ibid., 722b.
8. Ibid., 723a.
9. Ibid., 542b.
10. Ibid, 542a.

Chapter 15. Synchronizing Cosmologies: The Na-khi-Dongba of China

1. Wheatley, *Pivot of the Four Quarters,* 423–27; Snodgrass, *Architecture, Time, and Eternity,* 331.
2. Wheatley, *Pivot of the Four Quarters,* 231.
3. Ibid., 225.
4. Mathieu, *Study of the Ancient Kingdom,* 8.
5. Rock, *Na-Khi–English Encyclopedic Dictionary,* 300.
6. Ibid., 298.
7. Ibid., 300.
8. Goullart, *Forgotten Kingdom,* 93.
9. Mathieu, *Study of the Ancient Kingdom,* 55.
10. Ibid, 55.
11. Mathieu, *Study of the Ancient Kingdom,* 55.
12. Rock, *Na-Khi–English Encyclopedic Dictionary,* 366.
13. Ibid., xxi and 365.
14. Mathieu, *Study of the Ancient Kingdom,* 148.
15. He, *Naxi Dongba Pictograph Dictionary,* 13.
16. Rock, *Na-Khi–English Encyclopedic Dictionary,* 49.

Chapter 16. As Above, So Below: The Chariot of Orion

1. Griaule and Dieterlen, *Pale Fox,* 500.
2. Budge, *Egyptian Hieroglyphic Dictionary,* 98b.
3. Griaule and Dieterlen, *Pale Fox,* 503.
4. Ibid., 504.
5. Scranton, *Sacred Symbols of the Dogon,* 129.
6. Budge, Egyptian Hieroglyphic Dictionary, 174b.
7. India Profile, "Of Golden Brilliance—Surya God of Light," www.indiaprofile.com/religion-culture/surya.htm (accessed April 27, 2010).

BIBLIOGRAPHY

Best, Elsdon. *Maori Religion and Mythology.* Wellington, New Zealand: W. A. G. Skinner, Government Printer, 1924.

Bon Future Fund Foundation. "About the Bon." www.bonfuturefund.org/wp/about-the-bon.

Budge, E. A. Wallis. *An Egyptian Hieroglyphic Dictionary.* New York: Dover Publications, Inc., 1978.

Calame-Griaule, Genevieve. *Dictionnaire Dogon.* Paris: Librarie C. Klincksieck, 1968.

Goullart, Peter. *Forgotten Kingdom.* London: Readers Union/John Murray, 1957.

Griaule, Marcel. *Conversations with Ogotemmeli.* Oxford: Oxford University Press, 1970.

Griaule, Marcel, and Dieterlen, Germaine. *The Pale Fox.* Paris: Continuum Foundation, 1986.

Hagan, Helene. *The Shining Ones: An Entymological Essay on the Amazigh Roots of Egyptian Civilization.* Philadelphia: Xlibris, 2000.

Hart, George. *A Dictionary of Egyptian Gods and Goddesses.* London and New York: Routledge, 1999.

He Pingzheng. *Naxi Dongba Pictograph Dictionary.* Translated by Xuan Qin. Lijiang, China: Yunnan Fine Arts Publishing House, 2004.

Long, Charles H. *Alpha: The Myths of Creation.* New York: George Braziller, 1963.

Lucas, J. Olumide. *The Religion of the Yorubas.* New York: Athelia Henrietta Press, 2001.

Mathieu, Christine. *A History and Anthropological Study of the Ancient

Kingdoms of the Sino-Tibetan Borderland—Naxi and Mosuo. Lewiston, Australia: The Edwin Mellen Press, 2003.

Milnor, Seaver Johnson. "A Comparison Between the Development of the Chinese Writing System and Dongba Pictographs." In *University of Washington Working Papers in Linguistics*. Vol. 24, edited by Daniel J. Jinguji and Steven Moran, 30–45. Seattle, Wash.: University of Washington Linguistics Department, 2005. http://depts.washington.edu/uwwpl/vol24/pub_submission_Milnor.pdf.

Renouf, Peter Le Page. *The Egyptian Book of the Dead: Translation and Commentary*. Privately printed by Edouard Naville for the Society of Biblical Archaeology, London, 1904.

Rock, Joseph. *A Na-Khi—English Encyclopedic Dictionary, Part I*. Rome: Istituto Italiano Per Il Medio Ed Estremo Oriente, 1963.

———. *The Ancient Na-Khi Kingdom of Southwest China*. Cambridge, Mass.: Harvard University Press, 1947.

Rosemont, Henry, Jr. *Explorations in Early Chinese Cosmology*. Chico, Calif.: Scholars Press, 2006.

Sagan, Carl. *Broca's Brain*. New York: Random House, 1979.

Scranton, Laird. "Revisiting Griaule's Dogon Cosmology: Comparative Cosmology Provides New Evidence to a Controversy." *Anthropology News*, April 2007, 24–25.

———. *Sacred Symbols of the Dogon: The Key to Advanced Science in the Ancient Egyptian Hieroglyphs*. Rochester, Vt.: Inner Traditions, 2007.

———. *The Science of the Dogon: A Study of the Founding Symbols of Civilization*. Rochester, Vt.: Inner Traditions, 2006.

Snodgrass, Adrian. *Architecture, Time and Eternity*. New Delhi, India: Aditya Prakashan, 1990.

———. *The Symbolism of the Stupa*. Delhi, India: Motilal Banarsidass Publishers, 1992.

Van Beek, Walter. *Dogon: Africa's People of the Cliffs*. New York: Harry N. Abrams, Inc., 2001.

Winters, Clyde A., "Proto-Saharan Religions." www.ipoaa.com/proto_saharan_religions.htm.

Wheatley, Paul. *The Pivot of the Four Quarters: A Preliminary Enquiry into the Origins and Character of the Ancient Chinese City*. Chicago: Aldine Publishing Company, 1971.

INDEX